Energy, Climate and the Environment Series

Series Editor: **David Elliott**, Emeritus Professor of Technology, Open University, UK

Titles include:

Manuela Achilles and Dana Elzey (*editors*)
ENVIRONMENTAL SUSTAINABILITY IN TRANSATLANTIC PERSPECTIVE
A Multidisciplinary Approach

Luca Anceschi and Jonathan Symons (*editors*)
ENERGY SECURITY IN THE ERA OF CLIMATE CHANGE
The Asia-Pacific Experience

Philip Andrews-Speed
THE GOVERNANCE OF ENERGY IN CHINA
Implications for Future Sustainability

Ian Bailey and Hugh Compston (*editors*)
FEELING THE HEAT
The Politics of Climate Policy in Rapidly Industrializing Countries

Gawdat Bahgat
ALTERNATIVE ENERGY IN THE MIDDLE EAST

Mehmet Efe Biresselioglu
EUROPEAN ENERGY SECURITY
Turkey's Future Role and Impact

David Elliott (*editor*)
NUCLEAR OR NOT?
Does Nuclear Power Have a Place in a Sustainable Future?

David Elliott (*editor*)
SUSTAINABLE ENERGY
Opportunities and Limitations

Horace Herring and Steve Sorrell (*editors*)
ENERGY EFFICIENCY AND SUSTAINABLE CONSUMPTION
The Rebound Effect

Horace Herring (*editor*)
LIVING IN A LOW-CARBON SOCIETY IN 2050

Matti Kojo and Tapio Litmanen (*editors*)
THE RENEWAL OF NUCLEAR POWER IN FINLAND

Antonio Marquina (*editor*)
GLOBAL WARMING AND CLIMATE CHANGE
Prospects and Policies in Asia and Europe

Catherine Mitchell, Jim Watson and Jessica Whiting (*editors*)
NEW CHALLENGES IN ENERGY SECURITY
The UK in a Multipolar World

Catherine Mitchell
THE POLITICAL ECONOMY OF SUSTAINABLE ENERGY

Ivan Scrase and Gordon MacKerron (*editors*)
ENERGY FOR THE FUTURE
A New Agenda

Gill Seyfang
SUSTAINABLE CONSUMPTION, COMMUNITY ACTION AND THE NEW ECONOMICS
Seeds of Change

Benjamin K. Sovacool
ENERGY & ETHICS
Justice and the Global Energy Challenge

Joseph Szarka
WIND POWER IN EUROPE
Politics, Business and Society

Joseph Szarka, Richard Cowell, Geraint Ellis, Peter A. Strachan and Charles Warren (*editors*)
LEARNING FROM WIND POWER
Governance, Societal and Policy Perspectives on Sustainable Energy

David Toke
ECOLOGICAL MODERNISATION AND RENEWABLE ENERGY

Thijs Van de Graaf
THE POLITICS AND INSTITUTIONS OF GLOBAL ENERGY GOVERNANCE

Xu Yi-chong (*editor*)
NUCLEAR ENERGY DEVELOPMENT IN ASIA
Problems and Prospects

Xu Yi-chong
THE POLITICS OF NUCLEAR ENERGY IN CHINA

Energy, Climate and the Environment
Series Standing Order ISBN 978–0–230–00800–7 (hb)
978–0–230–22150–5 (pb)

You can receive future titles in this series as they are published by placing a standing order. Please contact your bookseller or, in case of difficulty, write to us at the address below with your name and address, the title of the series and the ISBNs quoted above.

Customer Services Department, Macmillan Distribution Ltd, Houndmills, Basingstoke, Hampshire RG21 6XS, England

Environmental Sustainability in Transatlantic Perspective

A Multidisciplinary Approach

Edited by

Manuela Achilles
Lecturer and Associate Director, Centre for German Studies, University of Virginia, USA

and

Dana Elzey
Associate Professor, University of Virginia, USA

palgrave
macmillan

First published 2013 by
PALGRAVE MACMILLAN

Palgrave Macmillan in the UK is an imprint of Macmillan Publishers Limited, registered in England, company number 785998, of Houndmills, Basingstoke, Hampshire RG21 6XS.

Palgrave Macmillan in the US is a division of St Martin's Press LLC, 175 Fifth Avenue, New York, NY 10010.

Palgrave Macmillan is the global academic imprint of the above companies and has companies and representatives throughout the world.

Palgrave® and Macmillan® are registered trademarks in the United States, the United Kingdom, Europe and other countries.

ISBN 978–1–137–33447–3

This book is printed on paper suitable for recycling and made from fully managed and sustained forest sources. Logging, pulping and manufacturing processes are expected to conform to the environmental regulations of the country of origin.

A catalogue record for this book is available from the British Library.

A catalog record for this book is available from the Library of Congress.

Typeset by MPS Limited, Chennai, India.

We dedicate this book to our children, Yakob and Hannah, and Anna, Claudia and Ethan. The editors also gratefully acknowledge the support of the Center for German Studies at the University of Virginia, as well as of Dorothe Bach, Associate Director of the UVa Teaching Research Center.

Contents

List of Tables and Figures

Tables

Figures

Series Editor Preface

Energy, Climate and the Environment

Concerns about the potential environmental, social and economic impacts of climate change have led to a major international debate over what could and should be done to reduce emissions of greenhouse gases. There is still a scientific debate over the likely *scale* of climate change and the complex interactions between human activities and climate systems, but global average temperatures have risen and the cause is almost certainly the observed build up of atmospheric greenhouse gases.

Whatever we now do, there will have to be a lot of social and economic adaptation to climate change – preparing for increased flooding and other climate-related problems. However, the more fundamental response is to try to reduce or avoid the human activities that are causing climate change. That means, primarily, trying to reduce or eliminate emission of greenhouse gasses from the combustion of fossil fuels. Given that around 80% of the energy used in the world at present comes from these sources, this will be a major technological, economic and political undertaking. It will involve reducing demand for energy (via lifestyle choice changes – and policies enabling such choices to be made), producing and using whatever energy we still need more efficiently (getting more from less) and supplying the reduced amount of energy from non-fossil sources (basically switching over to renewables and/or nuclear power).

Each of these options opens up a range of social, economic and environmental issues. Industrial society and modern consumer cultures have been based on the ever-expanding use of fossil fuels, so the changes required will inevitably be challenging. Perhaps equally inevitable are disagreements and conflicts over the merits and demerits of the various options and in relation to strategies and policies for pursuing them. These conflicts and associated debates sometimes concern technical issues, but there are usually also underlying political and ideological commitments and agendas which shape, or at least color, the ostensibly technical debates. In particular, at times, technical assertions can be used to buttress specific policy frameworks in ways which subsequently prove to be flawed.

The aim of this series is to provide texts which lay out the technical, environmental and political issues relating to the various proposed policies for responding to climate change. The focus is not primarily on the science of climate change or on the technological detail, although there will be accounts of the state-of-the-art, to aid assessment of the viability of the various options. However, the main focus is the policy conflicts over which strategy to pursue. The series adopts a critical approach and attempts to identify flaws in emerging policies, propositions and assertions. In particular, it seeks to illuminate counter-intuitive assessments, conclusions and new perspectives. The aim is not simply to map the debates, but to explore their structure, their underlying assumptions and their limitations. Texts are incisive and authoritative sources of critical analysis and commentary, indicating clearly the divergent views that have emerged and also identifying the shortcomings of these views. However, the books do not simply provide an overview, but they also offer policy prescriptions.

The present text brings together a range of authors from the USA and Germany to explore the commonalities and differences in the way in which the technological, social, economic and cultural changes associated with the need to respond to climate change are perceived and pursued in their respective countries. The starting points of these two countries are very different. Germany has a major political commitment to technological change, driven by climate concerns and also by widespread hostility to nuclear power. The USA has shown much less commitment to responding to climate change, but its market-based system, augmented by federal stimulus funding, has resulted in a growth in the deployment of renewable energy technologies on almost the same scale, proportionately, as in Germany. That said, the wider political differences are very large. Germany has a major influence over EU policies, but the USA, as a global superpower, has massive political, economic and of course military influence internationally.

Given this broad context, the chapters in this book map out both differences and similarities in underlying aims, values and assumptions, as well as practices and policies, at a range of levels, local, national and international. Mutual understanding is important even when there are differences, but as this book indicates, there are some convergent trends and a potential for beneficial cross-Atlantic co-operation and collaboration.

Preface and Acknowledgments

The 21st century has been referred to as the 'up or out' century, implying that we will approach a tipping point with respect to the increasing burden we can place on the Earth's resources and ecosystem within the next two to three generations. No comprehensive and authoritative work exists that has answers to the questions that will define this century: how many people and at what quality and standard of living can the finite resources and regenerative capacities of the planet sustain, and for how long? Though some have been raising the specter of resource exhaustion, environmental degradation and climate change for decades, we have really only just begun to grapple with the enormity and complexity of these challenges.

Just as we, as individual human beings, are different and respond in different ways to a given challenge, so human societies are different and will respond to these global challenges of the 21st century differently. This is a positive for two reasons: firstly the response must be carefully designed to suit the values, capabilities and circumstances of each individual society, and secondly, a variety of responses affords us the opportunity to observe and learn from what works, and doesn't work, as societies develop and implement particular approaches. The aim of this book is exactly this, to present different views and responses to these global challenges, with a specific focus on the United States and Germany; it is thus a comparative, transatlantic exploration of environmental sustainability.

This transatlantic perspective is rich with both contrast and parallel. The transatlantic bonds go deep, with roughly one of every six people in the United States having German ancestry. English is a Germanic language. German Americans make up roughly one-third of the German diaspora worldwide, and Germany is still among the top destinations for US students studying abroad. Both Germany and the United States have consistently challenged convention and pioneered new technologies and ways of doing things, from education and transportation to organization and communication. Both have reputations for their work ethic, research and engineering, and level of organization. Yet the two nations' responses to climate change and sustainability are remarkably different, primarily in the sense that the United States has not yet taken a serious, national stance, one that affects ordinary American citizens in the same way that

Germans are affected in their daily lives by the decisions German government and citizens groups have made at every level. It might be that the US response would not be so different from that of Germany, if both were being developed and implemented at the same rate and scale.

One of the many insights this book successfully highlights is the scale and boldness with which Germans are preparing to deal with the 'gathering storm'. Is this 'energy transition' (*Energiewende*) an appropriate, informed response, or are Germans the 'doomsday preppers' on a national scale? Is the United States' cautious reluctance to fully embrace sustainability as a national challenge the wiser path, or is the political paralysis that has gripped Washington DC placing Americans dangerously (and perhaps the rest of the planet) at risk?

These questions bring us back to a central point of this book, which is that national responses, both with regard to their nature and extent, are reflections of more deeply held values, which in turn are products of historical experience, cultural factors and demography. Whether the disciplinary perspective is that of architecture, economics, statecraft or biology, analysis and interpretation of the response, as well as the development of any new, proposed solutions, must rest on an understanding, and sensitivity to, these contextual factors. What is more, each of us must find the opportunities and means to share the knowledge and practice from within our own disciplines, while being open to appreciation and integration of the work of other disciplines.

Bringing very different disciplines together to present analyses and perspectives on a given topic inevitably entails challenges, but genuine efforts at solution must reflect the multi-dimensional character of the problems we face. Assessing the impacts of human decisions and actions on the environment, as well as the efforts to address the consequences, encompasses the built environment, the natural world and the ongoing negotiation between the two; it therefore embraces an enormous range of human endeavor, of science and technology. It has been our pleasure to work with the various contributors to this book and, with their unique insights and experience, to share this transatlantic perspective on what is perhaps the most important question of our time. As both contributors and editors, we have learned a great deal from them. We are confident that you will as well.

Notes on Contributors

Manuela Achilles is Associate Director of the Center for German Studies at the University of Virginia (UVa) (since 2008) and teaches in the History Department and Department of German Studies (since 2006). She holds a PhD in history and German studies from the University of Michigan, and has published on topics at the intersections of history and literature, with a focus on modern Germany.

Timothy Beatley is Teresa Heinz Professor of Sustainable Communities, Urban and Environmental Planning at UVa. He holds a PhD from the University of North Carolina at Chapel Hill. He is an expert in environmental planning and policy and the author of numerous books, including *Green Urbanism: Learning from European Cities*. A recent documentary is titled *The Nature of Cities*.

Peter Debaere is Associate Professor in the Darden Graduate School of Business at the UVa, and research fellow at the Center for Economic Policy Research. Before coming to UVa in 2006, he held a position at the University of Texas, Austin. Peter Debaere has a PhD from the University of Michigan and is an expert on globalization and international trade. His most recent research focuses on the global economics of water.

Dana Elzey is Associate Professor of Materials Science and Engineering at UVa. In addition to teaching and research, he is director of the Rodman Scholars (honors engineering) Program and of International Programs for the School of Engineering and Applied Science. He holds a doctorate from the Universität Stuttgart, joining UVa in 1990. His areas of expertise are design thinking, international engineering programs and the processing, properties and mechanics of advanced composites.

Arne Jungjohann heads the Department for Political Strategy in the State Ministry of Baden-Württemberg. Previously, he directed the Environment and Global Dialogue Program of the Heinrich Böll Foundation in Washington DC and a senior advisor in the German Bundestag.

Carmen Kristan has been policy adviser for climate and energy at the German Embassy in Washington DC since September 2011. The main focus of her work is the Transatlantic Climate Bridge (www.germany.info/climatebridge). Previous to her DC work, she worked on the Transatlantic

Climate Bridge in the Southeast of the United States with the German Consulate General where she had a position as an Advisor for Public Affairs since 1999.

Anja Kueppers-McKinnon obtained her MA in journalism in Sydney, Australia. Upon her return to Europe, she worked as a radio reporter, producer and presenter for *Deutsche Welle* Radio in Bonn, Germany. In 2008, Anja moved to Washington DC where she took on the role of Policy Advisor for Climate and Energy at the German embassy. Since 2011, she has been living back in her home town of London where she works as a freelance radio and print journalist.

Brian Marrs holds a degree in Foreign Affairs from UVa and a master's degree from the Yale School of Forestry and Environmental Studies in energy economics and policy. Prior to Yale, Brian served as a Robert Bosch Foundation Fellow at the German Energy Agency (dena) and Vattenfall Europe. He received awards from the Yale Entrepreneurial Institute and a Sobotka Venture Prize for launching an energy-focused start-up company. Brian served as a contributing writer to *Corporation 2020*, a book edited by Pavan Sukhdev, former Special Envoy and Goodwill Ambassador to the United Nation's Environmental Program.

Georg Maue serves as First Secretary for Climate and Energy at the German embassy in Washington DC. Prior to joining the embassy, Georg worked at the German Federal Ministry for the Environment, Nature Conservation, and Nuclear Safety, with responsibility for climate and energy policy and various international environmental topics at the German Environmental Protection Agency. He collaborated on the ministry's development of recent energy policies targeted at moving the German energy system in a cost-effective manner toward greater use of renewable energies and higher efficiency. He holds an engineering degree in environmental technologies and received his PhD in microbiological treatment of contaminated soils from the Technical University of Berlin.

Dale Medearis is a senior environmental planner for the Northern Virginia Regional Commission (NVRC) and is an adjunct professor at Georgetown University. He co-leads the NVRC's regional climate mitigation and energy programs and manages NVRC's international environmental partnerships. Prior to working for NVRC, he worked at the Office of International Affairs, US Environmental Protection Agency, Washington DC, as the program manager for Western Europe and urban environmental programs. He has been the Vice-chair of the OECD Territorial Development Committee and

Chairman of the OECD Working Group on Urban Affairs. He has a PhD in environmental design and planning from Virginia Tech University.

Michael Mehling is President of the Ecologic Institute in Washington DC. Previously, he has held academic appointments at the Universities of Konstanz, Georgetown, Greifswald and Helsinki. He is a contributing author to the Fifth Assessment Report of the Intergovernmental Panel on Climate Change (IPCC).

Eugene Ryang is a landscape architect and co-founding principal of Waterstreet Studio (a landscape architecture, civil engineering and conservation planning firm) in Charlottesville, Virginia. He received master's degrees in landscape architecture and cultural anthropology from UVa and a BA in History from Columbia University. He leads multi-disciplinary teams on public and private commissions ranging from residential and farm planning projects to schools and universities; parks and arboretums; and eco-retreats/resorts and conservation land developments throughout the United States.

Jörg Sieweke is Assistant Professor in Landscape Architecture at UVa. Before coming to UVa in 2009, he held positions at the Academy of Arts in Berlin and Stuttgart and at the Technical University in Dresden. He is a licensed urban designer and landscape architect in Germany.

Kerstin Steitz is a doctoral candidate in the Department of Germanic Languages and Literature at University of Virginia. Currently, she is finishing her dissertation entitled 'Beyond Closure: The Artistic Re-Openings of Holocaust Trials'. Since 2009, she has served as teaching and program assistant for several short-term study abroad programs at UVa, including *Global Ingenuity 21* and *Global Technology Practice*.

Mark White is Associate Professor at the UVa's McIntire School of Commerce and Visiting Professor of Environmental Economics at the Technische Universität Dresden. He holds advanced degrees in ecology, business and finance from Michigan State University. His academic expertise is in the fields of corporate finance and sustainable business practices.

Introduction: Environmental Sustainability in Transatlantic and Multidisciplinary Perspective

Manuela Achilles and Dana Elzey

Building a sustainable society is a major challenge of the 21st century. Striving to become the world's first major renewable energy economy by 2050, Germany is widely considered a global front-runner in environmental policy and practice. Requiring large investments in green technologies, as well as in new power lines and energy storage systems, the German shift from fossil fuels and nuclear power towards renewables amounts to a veritable 'energy revolution'. What are the challenges and opportunities of this transition toward a more sustainable future? How did Europe's largest economy come to embrace an energy challenge that has been compared to the first landing on the moon? And most importantly, is the German experience transferable to other industrialized nations such as the United States?

In this volume, leading experts from the business world, academia, governmental agencies and non-profit think tanks approach key questions of environmental sustainability from a transnational and multidisciplinary angle. Based on an invited lecture series hosted by the Center for German Studies at the University of Virginia in 2009, the carefully calibrated chapters open up new perspectives on environmental sustainability at all levels of governance and on the possibilities for transatlantic partnership and cooperation. Covering a large area of expertise, one of the authors' major objectives is to generate a productive dialogue across academic disciplines, and between theory and practice. Sustainability, this volume suggests, is not just a question of engineering, architecture, politics or economics. Real-world solutions require a more integrated approach.

Expanding its reach from the global to the national and local level, the broader theme of this book concerns the creation and use of energy and the societal consequences of the environmental choices we make for our future. The continued depletion of finite fossil fuel resources; degradation of our air, soil and water; and climate change all threaten human progress and perhaps even our existence. How we respond to these challenges at the level of societies is determined by a set of historical, socio-cultural, demographic and other factors, unique to each society. The actions a society takes may be in the form of new policies and regulations, economic reforms, political re-organization or technological innovation. Figure I.1 illustrates, in simplified schematic form, this circle: human action (e.g. creation and use of energy), consequence, filtration of the impacts through the lens of our historical and present context, and response. The response, that is, the action taken in responding to the imposed environmental challenges, is a reflection of the context of human life, unique to each society.

Another important and equally broad theme of this volume concerns our use of technology. The application of our scientific understanding of the natural world, to create a human-built, engineered world, is an incredibly powerful tool. We have the intelligence to engineer the world around us, but do we have the ability to manage its use responsibly? Our record thus far is not promising, but there are bright spots. The context of human life, comprising historical, cultural, political and other factors, may be thought of as a process in which observed changes in

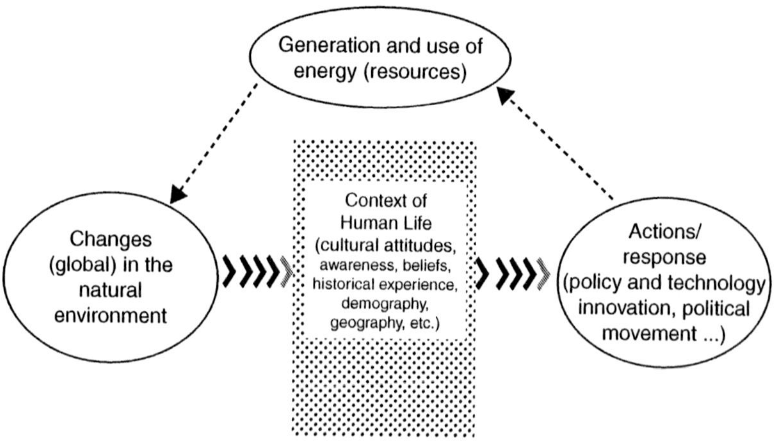

Figure I.1 Human action – environmental response feedback loop
Note: The context of human life, unique to each human society, pictured as a process for translating observations (data) about the environment, resource availability and so on into actions. These actions, when implemented, affect the use of energy and resources, and thus a further change in the environment.

the world around us are interpreted and translated into action (again, refer to Figure I.1); as this context is unique to each individual society, so the process of recognizing, prioritizing and addressing challenges is culturally unique.

The fact that we cannot interpret and respond to global challenges, except through the unique, individual lens of our own socio-cultural *Zeitgeist*, has multiple implications. Firstly, it suggests that we may expect a wide range of alternative responses to the same challenges or threats and that what may prove successful in one context may not be so in another. But while there may be no 'one size fits all' solution, we should seek to observe and understand the responses of others and to identify what may be considered 'best practices', and to coordinate efforts wherever possible. A further implication of this contextual lens is that, occasionally, there will be responses that appear unsound or even irrational to outside observers. Deeper knowledge of the contextual environment in which these decisions are made would, however, render them conceivable, if not rational. This points to the critical role of cross-cultural, social, political and historical understanding when seeking to negotiate compromise and implement multinational agreements on energy, resources and the environment.

We should also recognize, explicitly, that the contextual lens through which we view the world is changing with time. One significant impact on the context of human life for virtually everyone on the planet is globalization. As we evolve from a world of isolated, individual nations and societies, to a much more globally interconnected world, cultural and other differences among societies tend to become increasingly mitigated. Internet, high-speed communications networks, globalization of markets, economic interdependence and the challenges presented by global risks or threats have already contributed significantly to this process. This would suggest a growing level of comfort and understanding among individual nations in the future and an improved ability to achieve a higher degree of understanding and mediated consensus in responding to global challenges. This book may be read as a contribution to this end.

The first three chapters frame the broader context of sustainable development. In Chapter 1, Mark White investigates the ways in which we routinely treat the environment as a free public good without a price. Nature provides humanity with goods and services critical for our well-being, but only a few (e.g. food, timber, water) have been priced and are traded in markets. Drawing on the work of German systems thinker Frederic Vester, White introduces the notion of 'ecosystem services' and discusses methods economists use to value these services, together with the likelihood of continued supply under various future scenarios. Valuation of environmental amenities is of high concern, especially for the citizens of developed economies such as Germany or the United States, because prices reflect scarcity. Without complete information about the relative scarcity of an asset (as when assets remain unpriced) economic allocations are suboptimal and inefficient. White argues that 'putting a price on the planet' will help us to better assess the consequences of our consumptive desires and behaviors. As population and consumption expand and accessible natural resources continue to decline, the need for valuation and markets will only increase.

In Chapter 2, Peter Debaere offers a global economic perspective on environmental policies. In particular, he investigates the contentious pollution haven hypothesis, which has been at the forefront of debates about globalization since the riots surrounding the WTO meetings at Seattle in 1999. The hypothesis also offers a framework for thinking about potential impacts of fighting climate change on international trade. It predicts that free trade will make advanced economies with high environmental standards off-shore the production of pollution-intensive goods to less developed countries with lower environmental standards. Debaere argues that trade liberalizations *alone* should not shift the production of the dirtier goods to less advanced economies. In fact, the opposite should happen, since advanced economies such as Germany and the United States have a comparative advantage in the most polluting industries, such as chemicals, iron and steel. It is the asymmetric raising of environmental standards (and, by extension, carbon emission standards) in advanced economies that may trigger off-shoring of polluting industries to less developed countries with lower standards. To address this environmental concern, Debaere points to the role of technological progress in pollution reduction (reduction in carbon emissions) and the benefits of technology transfer to less developed countries. The chapter ends with a call for more research into the causes of technological innovation, including the role of governments and regulation. In light of the different policy approaches in advanced countries such as Germany and the United States, Debaere wonders whether Germany's more pro-active policies will, indeed, make it the preferred location for environmentally friendly industries.

Zooming in on how and where we live, Chapter 3 by Timothy Beatley explores the accelerated urbanization in much of the world. Can the rapid shift to cities be respectful and even restorative of nature? How can urbanites be challenged to see themselves as responsible ecological citizens? Beatley draws on urban design examples from virtually all continents to highlight patterns of green city planning that harness the positives of urban living while at the same time reducing its ecological footprint. Particular attention is given to pedestrian-friendly, compact cities that produce food as well as electricity more locally, while also enhancing the quality of life. The concept of the 'glocal' is used to situate cities below and beyond the radar of the national state. One objective of green city building, Beatley asserts, is to facilitate a combination of local and global connectedness that diminishes rather than increases the tremendous resource pressures on the planet. The wealth of examples discussed in the chapter indicates that there is already much happening. The challenge, Beatley suggests, is to find the political will and economic resources to pursue the new urban designs that have already been tested across the world.

Chapters 4 to 7 draw the contours of an environmental policy that is quite different from conditions in the United States, while offering many points of inspiration and contact. Germany provides a particularly relevant example for readers in the United States in demonstrating that a highly developed country can be environmentally conscious and at the same time competitive in the global market. Chapter 4 by Arne Jungjohann traces the transformation of the German Greens from an oppositional and heterogeneous social movement into a broadly based and goal-oriented people's party willing and capable of forming effective governments with different political partners at both the regional and federal level. The chapter provides an overview of the party's history and its impact on the country's environmental policies. Of particular interest are the Environmental Tax Reform, the Renewable Energy Act and the nuclear phase-out. The following chapters explore those policies in more detail, thus shifting the focus from the remarkable success story of the Greens to the broader political and economic dynamics of the German switch from fossil fuels and nuclear power to renewable energies.

Chapter 5 by Michael Mehling takes a closer look at the Ecological Tax Reform, one of the most debated factors in Germany's progression toward a sustainable economy. Recalling the warning from industry circles that the tax reform would stifle economic growth and prosperity, Mehling shows that the revenues were almost fully returned to taxpayers, with the largest share used for a gradual reduction of social security contributions. Rendering hiring less expensive, the environmental tax reform has contributed to the creation of jobs and hence strengthened the Germany economy. What is more, the targeted increase in energy costs encouraged the deployment of energy-efficient technologies and processes, including alternative energy sources. In Mehling's estimate – and in partial answer to Debaere's question – the greening of the German economy is unmistakably also the product of targeted policy design and implementation, including the Ecological Tax Reform. Energy-efficient technologies, Mehling notes, are now among the fastest growing German export products. As the incentive to reduce energy use also helped to make the German economy more resilient to fluctuations in global oil and gas prices, Mehling's balance sheet for the Ecological Tax Reform is clearly positive.

In Chapter 6, Manuela Achilles situates the German nuclear exit within both global and national contexts. Why is such a highly developed nation abandoning a technology that its supporters describe as carbon-neutral, plentiful and cheap? Achilles argues that it is hard to understand the German nuclear exit without reference to the country's history, politics and culture. Germany's material and moral devastation in two world wars, as well as the divided nation's exposed position within the cold war system of alliances, positioned the country in critical distance to nuclear power. Barred from the possession of the ultimate cold war weapon, Germany shed the militarism of the past. It then underwent a series of structural, social and political transformations that were driven by a green grassroots movement that rejected both the military and commercial use of nuclear power. In tracing this development from the Fukushima nuclear disaster back to the world historical conjuncture of 1945, Achilles opens a window onto a discussion of sustainability that is sensitive to both the global connectedness of environmental questions *and* the cultural determinants of the national response.

Chapter 7 by Brian Marrs offers a general industry perspective on Germany's switch from nuclear and fossil-fired energy sources to renewable energy. He discusses the German energy policies within their larger economic contexts, and assesses their practicality and costs. With considerable reservation, Marrs finds that the German energy revolution is technically achievable and can be expected to facilitate considerable job creation in the clean energy sector. He cautions, however, that the German energy switch will require investments of political, financial and technological capital on an unprecedented scale. Marrs also notes that current policy may contain troubling financial architecture, as the regulatory instruments designed to foster clean tech contribute to a rise in energy costs. In other words, the *Energiewende* (energy transformation) is not certain, and must continue to evolve over the coming decades to succeed. If Germany cannot retool its economy to produce clean, reliable and affordable energy, there will be few other countries with the political will and economic resources to do so.

The concluding five chapters explore different ways of cooperation and cross-fertilization between the United States and Germany, ranging from macro-level policies to micro-level projects and initiatives. Chapter 8 by Anja Kueppers-McKinnon, Georg Maue and Carmen Kristan presents the framework of the Transatlantic Climate Bridge (TCB). Launched by the German government in 2008, the TCB supports networks and partnerships that help Americans and Germans exchange their know-how and pave the way for joint solutions in the climate and energy arenas. The chapter discusses the TCB's activities at the local, state and federal level. Examples include practical exchanges on technological innovations and common standards, as well as the facilitation of innovative partnerships such as at the MIT – Fraunhofer Institute for Sustainable Energy located adjacent to the MIT campus in Cambridge, Massachusetts.

In Chapter 9, Dale Medearis discusses the transfer and application of environmental policies and lessons from Germany to the United States. His chapter starts with the observation that American environmental and energy policy has tended to be insular and introspective, lacking a global perspective that draws on international best practice. Against this background, Medearis reviews the initiatives of the Northern Virginia Regional Commission (NVRC), which has worked to develop and sustain a systematic study and testing of innovative, German sustainable development policies in the United States. Over the past ten years, Medearis notes, the NVRC has helped to co-launch the US/German Transatlantic Climate Bridge Initiative, created the first formal agreement on climate and energy between the US and European regional councils, launched the 4 Mile Run Watershed Restoration Project, and created two bilateral agreements between the Commonwealth of Virginia and the German environment and transportation ministries. Medearis stresses the investment and study within problem-focused, goal-oriented and geographical-specific contexts required for the transfer of experiences and innovations from one country to another. The NVRC has become a US model in this context. As the chapters by Anja Kueppers-McKinnon, Georg Maue and Carmen Kristan; Eugene Ryang; Timothy Beatley; and other contributors to this volume also show, trans-national and cross-cultural learning and exchange may well be happening at the local level, below and beyond the sensors of national states.

Chapter 10 by Dana Elzey and Kerstin Steitz outlines and analyzes two study abroad programs, designed to develop cross-cultural design thinking in engineering students at the University of Virginia (Charlottesville, Virginia). Though quite distinct in organization and approach, both programs revolve around the key insight that technological innovation is fundamentally a cultural activity. The authors report that participants come to understand sustainability and renewable energy not merely as areas of advanced technology but also as reflections of deeply held cultural values and attitudes. In relating this observation to what they know about engineering in the United States, American students become more aware of the cultural mediation of their own education and career paths. This leads to increased intercultural awareness and enhancement of specific problem-solving skills essential for professional engineers in a globalized environment.

The final two chapters discuss innovative design projects that internalize the transatlantic perspective. The chapters correlate work by Jörg Sieweke, a German urban designer who teaches architecture at the University of Virginia in the United States, and by American landscape architect Eugene Ryang, who adapts and applies German design ideas and concepts in Virginia. In Chapter 11, Jörg Sieweke explores novel ways in which landscape architects are dealing with highly contaminated post-industrial landscapes. His proposal for the remediation of a brownfield site on the Rhine river in Duisburg, Germany, exemplifies the paradigmatic shift from the eradication of the highly problematic industrial legacy toward its incorporation and reinterpretation. The challenge Sieweke faced in the RhinePark project was to reference the industrial heritage in an archeological sense, as all traces of the difficult past were originally planned to be buried underground. Sieweke's design deals with the site's considerable soil contamination by employing three different concepts of green. A generic lawn is established by putting a layer of fertile earth atop an impervious PVC seal. This conventional blanket approach is contrasted with the recovery of fifty acres of wide open meadow. Nature reestablished is claimed by relocating the top two meters of contaminated soil to a capped landfill that purposefully displays its artificial surface. Sieweke asserts that the three dimensions of green correspond with both the park's history and its mixed use by an increasingly diverse population.

Chapter 12 by Eugene Ryang describes an eco-revelatory landscape design for a bank site in Charlottesville, Virginia. Located in a historic college town at the scenic foothills of the Blue Ridge Mountains, this practice-oriented project presents an interesting contrast to Sieweke's theory-driven reclamation of a highly polluted industrial site in densely populated, central Europe. The underlying ideas are closer than one might gather on first sight, however. A common reference point is German architect Peter Latz, whose path-breaking contextual design projects reject a tabula rasa approach that imposes new structures on sites commonly understood as blank slates. Sieweke and Ryang both agree with Latz's suggestion to incorporate rather than erase given structures, but their tasks and perspectives are different. Sieweke's design is a work of remembrance that inscribes the history of man-made pollution into the reconstructed setting of a public park. Ryang and his team, on the other hand, seek to reveal and restore the existing hydrologies and ecologies of an urban infill site. His particular concern is with a damaged urban stream that most conventional design approaches would force underground. Drawing inspiration from German landscape architect Herbert Dreiseitl, whose regenerative waterscapes incorporate art and design with sustainable rainwater and stormwater management, Ryang and his team recover the biodynamic operation of the local stream in an aesthetically compelling and environmentally sensitive manner. After convincing the clients not to bury the stream in a box culvert (as is often the case with urban infill development), the team focused their efforts on the restoration of the impaired waterway. They implemented water quality swales and nutrient uptake cells to capture, hold and filter pollutants from the site's impervious surfaces (i.e. building roofs and parking lots) and to slowly release the 'clean' water back into the stream. These strategies are particularly important as the health of the Chesapeake Bay and its watershed, within which the bank site sits, has become a hot button issue in recent years. Final cost analyses (before implementation) showed this approach to be less expensive than burying the stream and paving over it. Interestingly, the bank president revealed that bank customers switched their accounts over to the new branch because they appreciated the site experience – a testament to how eco-revelatory approaches to site development can be both environmentally and economically regenerative.

Figure I.2, which presents a more detailed or 'magnified' view of Figure I.1, allows us to place the chapters comprising this book into the framework described by the diagram. Closer inspection reveals that almost all authors address subjects focused on the action/response component of the cycle. This fact underscores the emphasis on 'doing' as opposed to merely studying or understanding. The chapters address a wide range of responses, ranging from political movements to industry re-organization to policies designed to incentivize the use of sustainable practices. But the authors, in focusing on transatlantic perspectives, also recognize the uniqueness of the German and American perspectives on energy,

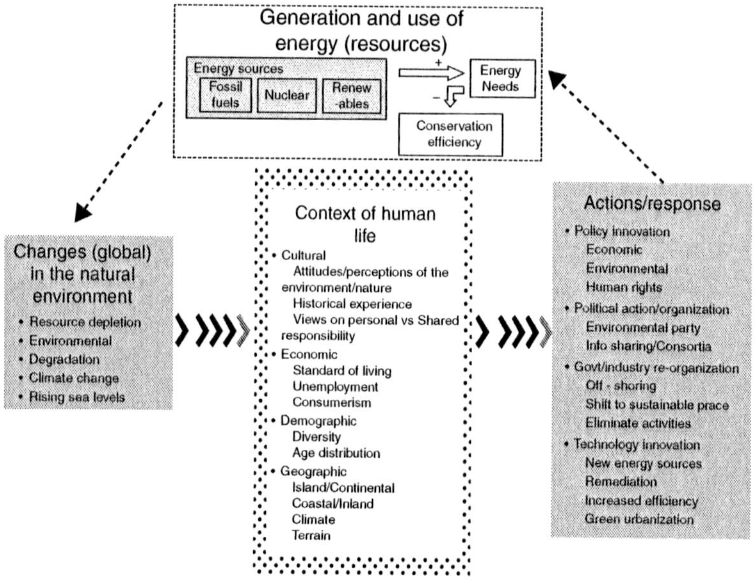

Figure I.2 Detailed human action – environmental response feedback loop
Note: This figure is a more detailed presentation of the feedback loop of Figure I.1; the focus of the chapters collected for this volume lie in the 'actions/response' area of the diagram, and therefore deal predominantly with assessment of current approaches, identification of best practices and on the applicability of responses to Germany and the United States.

the environment, sustainability and effective responses to global risks and challenges. They recognize the value in sharing these perspectives and in working together toward common ground and collaborative enterprise.

The chapters may be regarded as individually valuable contributions to understanding sustainability-related issues in their given field, but it is in reading the chapters across the spectrum of discipline and scale that the broader perspective, the themes mentioned above, truly begins to emerge. We wish you, the reader, an enjoyable and rewarding exploration of the perspectives and issues we have only been successful in bringing together collectively, as scholars, scientists and engineers, as economists, architects, historians, sociologists and environmental researchers. No one of us, individually, could have accomplished what you hold in your hand, perhaps a fitting metaphor for the challenges that lie ahead.

Part I
Broader Contexts of Sustainable Development

1
Putting a Price on the Plant: Economic Valuation of Nature's Services

Mark White

Artificial trees. What will they think of next?

The industrial revolution facilitated an enormous increase in human well-being – but with some unfortunate consequences. The burning of fossil fuels (coal, oil and gas) releases carbon dioxide (CO_2) into the atmosphere, where it accumulates and alters climate patterns. Atmospheric carbon dioxide concentrations are now around 390 ppm, more than 100 ppm higher than pre-industrial levels, and scientists believe this will lead to more intense storms, rising sea levels, increased flooding and droughts, and altered agricultural patterns as regions grow wetter or drier due to shifting patterns of rainfall.

Prior to the industrial revolution, carbon dioxide levels remained relatively constant. Animals inhaled oxygen and exhaled CO_2, while trees and other plants absorbed carbon dioxide and emitted oxygen. As human populations have increased, they've cut down trees and burned greater and greater amounts of fossil fuels – this is where the artificial trees come in.

Klaus Lackner, a physicist at Columbia University, has designed carbon dioxide absorption systems – aka artificial trees – that capture CO_2 a thousand times faster than real ones. It's clear they provide real benefits – allowing atmospheric carbon dioxide levels to increase unchecked is a recipe for climatic disaster – but at $30,000 apiece, the trees aren't cheap.

The observation that nature provides goods and services of benefit to humankind is not new; she's supplied humanity with food, fuel and fibers for millennia. Only in recent years, however, have we begun to understand and quantify the indirect benefits we receive from the

Acknowledgments: Mike Ellerbrock and Susie White provided helpful comments.

natural environment (e.g. carbon sequestration services, flood protection and climate regulation). In the coming years, as we struggle with increasing levels of population and consumption, and decreasing levels of fresh water, forests and biodiversity, we will see an increased need to assign economic values to nature's services. This chapter describes the challenges of 'putting a price on the planet': (1) identifying, defining and measuring specific ecosystem services, (2) valuing ecological amenities and (3) designing payment schemes to facilitate trade in nature's services.

The value of nature

Nature possesses intrinsic value (e.g. biodiversity) as well as anthropocentric benefits. Frederic Vester, a German biochemist and noted proponent of systems thinking (*vernetztes Denken*), was an early advocate for valuing nature in a holistic fashion. In *Der Wert eines Vogels* (1983), Vester observed that although the material value of a typical songbird is quite small, the services it provides to humanity are quite large. By 'material value', he meant the value a bird's respective chemical components (calcium, carbon, phosphorus, etc.) might fetch on the open market; he estimated this figure at 2¢. Vester then went on to enumerate the many benefits songbirds bestow on humankind, assigning economic values to each: insect removal ($36), tree planting ($12), environmental monitoring ($60), etc. until he came to a final value of $180. In arriving at these figures, Vester used prices for similar services, arguing, for instance, that the calming influence of a songbird's call equates to a year's supply of Valium ($25), an anti-anxiety drug. Although extremely well-known in Germany, Vester's work is relatively unfamiliar outside his native land.

In another notable attempt to raise awareness about the value of nature, Robert Costanza and his colleagues analysed more than 100 studies valuing 17 categories of services over a range of 16 types of ecosystems. Their results, published in the journal *Nature*, reported the total value of Earth's ecosystems to be on the order of $33 trillion – about twice the then-value of the gross domestic products of the world's nations (Costanza et al., 1997). Though criticized, this study represented a significant milestone in the valuation of nature's services.

The most ambitious attempt to inventory Earth's natural systems (although it did not assign monetary values) is the Millennium Ecosystem Assessment (MEA). Commissioned in 2000 by the United Nations Environment Program, the results were released in 2005 after a five-year study period involving 1,360 experts and more than 900 reviewers. The report focused on ecosystem services (i.e. services

Category	Description	Status
Provisioning services		
1. Food – Crops	Maize, wheat, soybeans, others	Enhancement
2. Food – Livestock	Cattle, swine, poultry	Enhancement
3. Food – Capture fisheries	Oceanic and freshwater	Decline
4. Aquaculture	Cultivation of fish, crustaceans and shellfish	Enhancement
5. Food – Wild foods	Tubers, berries, bushmeat, insects	Decline
6. Fiber – Timber	For construction use	Mixed
7. Fiber – Cotton, hemp, silk	For clothing and furnishings	Mixed
8. Fiber – Wood fuel	Fuelwood, charcoal	Decline
9. Genetic resources	Sources of diversity for biological evolution	Decline
10. Biochemicals	Natural medicines, pharmaceuticals	Decline
11. Fresh water	For drinking and agricultural purposes	Decline
Regulating services		
12. Air quality regulation	Removal of pollutants, oxygen regeneration	Decline
13. Climate regulation -- Global	Global water cycle and climate regulation	Enhancement
14. Climate regulation -- Local	E.g. wind breaks, coastal protection	Decline
15. Water regulation	Storage and release of intermittent flow	Mixed
16. Erosion regulation	Soil retention	Decline
17. Water purification/treatment	Watershed services including water quality	Decline
18. Disease regulation	Control of biotic disease vectors	Mixed
19. Pest regulation	Control of agricultural pests	Decline
20. Pollination	Crop fertilization services	Decline
21. Natural hazard regulation	E.g., flooding and storm mitigation	Decline
Cultural services		
22. Spiritual and religious values	Natural features with cultural and spiritual value	Rapid decline
23. Aesthetic values	Attractive landscape features	Decline
24. Recreation and ecotourism	Hiking, water sports, hunting, nature watching, etc.	Mixed
Supporting services	Primary production, nutrient cycling, soil formation	Mixed

Figure 1.1 Ecosystem services over the past 50 years
Source: Millennium Ecosystem Assessment (2005).

provided by natural ecosystems that contribute to human well-being). Also known as *natural capital*, the MEA categorized Earth's ecosystem services (Figure 1.1) as follows:

1. Provisioning services such as food, water, timber and fiber
2. Regulating services that affect climate, floods, disease, wastes and water quality

3. Cultural services that provide recreational, aesthetic and spiritual benefits and
4. Supporting services such as soil formation, photosynthesis and nutrient cycling.

The results of these efforts were sobering. We are spending Earth's natural capital and putting such strain on ecosystems that their ability to sustain future generations can no longer be taken for granted. Some of the report's key findings are as follows:

- About 60 percent (15 out of 24) of the ecosystem services identified in the report are being degraded or used unsustainably.
- There is established but incomplete evidence that the changes being made to ecosystems are increasing the likelihood of non-linear responses (i.e. accelerating, abrupt and potentially irreversible changes threatening human well-being).
- The harmful effects of ecosystem degradation are borne disproportionately by the poor and contribute to growing inequities and disparities.

Environmental degradation is already a significant barrier to achieving the Millennium Development Goals, a set of eight international development goals adopted by United Nations member states in 2000. Economic inequities are unsustainable for two reasons. First, they perpetuate poverty, which can lead to a vicious cycle involving increased degradation of environmental goods and services, which increases poverty, resulting in further degradation, and so on. Second, inequities hinder cooperation across different socio-economic classes, which will certainly be needed if we are to avert some of the more challenging problems involving common resources (e.g. freshwater, fisheries and the carbon-sequestering services of forests).

In addition to evaluating changes in the world's so-called ecological balance sheet, the authors of the MEA reported the results of a scenario planning exercise intended to provide insight into likely futures. Drawing upon various political, economic, social and technological trends, they advanced four possible scenarios:

Global orchestration

The closest to 'business as usual', this scenario imagines a globally connected society focused on global trade and economic liberalization.

Policymakers take a reactive approach to ecosystem problems. Economic growth is the highest, and population growth the lowest, in this scenario.

Order from strength

Sometimes referred to as the 'Mad Max' scenario for its bleak depiction of a regionalized and fragmented world, it is a future composed of primarily regional markets maintained through the use of military force. Little attention is paid to public goods like clean air, water and the like. Economic growth is the lowest and population growth the highest in this scenario.

Adapting mosaic

In the Adapting Mosaic scenario, regional watershed-scale ecosystems are the focus of political and economic activity. A proactive approach to ecosystem management is taken, resulting in greater resilience. This scenario is sometimes called the 'small is beautiful' scenario after economist E. F. Schumacher's influential book advocating such a path. Economic growth rates are lower than in the other scenarios but increase over time, and population is nearly as high as in Order from Strength.

Technogarden

This scenario imagines a highly managed, technologically laden future that recognizes the value of ecosystem services and either invents substitutes (artificial trees!) or requires payment for the supply of ecosystem services. It reflects a globally connected world relying on sound environmental technology and achieves relatively high levels of economic growth and population growth in the mid-range of the various scenarios. This scenario is sometimes known as 'Ecotopia', after Ernest Callenbach's 1975 utopian novel of the same name in which various earth-friendly technologies (e.g. homes made from extruded plant-based plastics) play important roles.

None of the MEA scenarios reflects a continuation of our present activities, although all are based on current conditions and trends. In some scenarios (e.g. Order from Strength), three of the four major categories of ecosystem services – provisioning services, regulating services and cultural services – experience declines. In others (e.g. Technogarden), some services show improvement (provisioning and regulating) while others (cultural) decline. Reversing the degradation of ecosystems in the face of increasing demand will require significant changes in policies, institutions and procedure, with a special emphasis on the identification and transfer of ecological benefits.

Well-established valuation systems and markets have existed for nature's provisioning services (crops, livestock, timber) for millennia. Until recently, other ecosystem services have been largely ignored because they weren't scarce and/or there were no markets for them. Coastal protection, climate regulation, soil formation and pollination services were just part of nature's backdrop. Today, increasing population and increasing levels of resource scarcity have changed all that. We now live in a 'full earth' and need an economic framework for this situation. Conventional economic decision-making won't work because many of our scarce environmental assets (e.g. clean air, soil retention) aren't traded in markets. One solution is to bring them into the market (e.g., through taxes and subsidies), or through cap-and-trade schemes. This is the realm of environmental economics, and we are making progress on this front, although nowhere near fast enough. The Netherlands' 'tap water tax' is an example of the former solution; the sulfur dioxide permit trading market authorized under the US Clean Air Act of 1990 exemplifies the latter.

Another approach is to recognize that we can't easily establish markets for all types of environmental goods and adjust our economic thinking (e.g. along the lines of an *ecological economics*).[1] Doing nothing is really not an attractive option. Because many of our environmental assets (oceanic fisheries, carbon sequestration services) are open access resources, inaction is likely to result in the tragedy of the commons – an unfortunate situation in which a resource is depleted because of individual, rational decisions. A key element in the valuation and establishment of markets for environmental amenities is the observation that many goods are public, rather than private goods.

This situation has been recognized by prominent economists in earlier times. The markets envisioned in Adam Smith's seminal treatise, *The Wealth of Nations* (1776), were concerned primarily with the provision of private goods, though they were couched within a society with strong moral and social constraints. E. F. Schumacher, writing in *Small is Beautiful* (1973), observed the impact of declining ecosystem services in India, but his prescriptions were largely ignored in the West as not relevant and/or too exotic. Herman Daly, former senior economist at the World Bank, wrote about the challenges of economic growth on a finite planet.[2] Although Daly was taken seriously, people did not really like the idea of uneconomic growth, or his suggested solution – a steady-state economics. In any event, valuation clearly lies at the heart of the problem.

Valuing nature

First of all, it is necessary to distinguish between value and prices. The aforementioned ecosystem services are clearly very valuable, but they are often unpriced. Adam Smith's diamond-water paradox illustrates the difference: Water is clearly more valuable to human society than diamonds, yet the latter trades at a higher price. Why?

The answer lies in the respective supply and demand for the two commodities. Because the supply of water is large relative to demand (at least until recently), water carries a low price. The demand for diamonds is large (at least relative to supply), so they are high-priced. Economics is concerned about prices, because prices signal scarcity and provide incentives for trade. Prices reflect what a good or service is worth to a marginal buyer (i.e. the next buyer to walk through the door). Hence, value lies at the margin. Utilitarian philosophers – individuals subscribing to the belief that one should seek the greatest good for the greatest number of people – assert that a product's or service's value to society is the sum of its value to individual members, and that prices provide a good estimate for this value.

Alas, as we have noted, many ecosystem services are not traded in markets, and hence, no prices can be observed. Economists, still intent on evaluating trade-offs between scarce, marketable resources and until recently, not-so-scarce nonmarketable resources, have developed a number of techniques for estimating the value of non-marketed environmental assets. In so doing, they've identified several different categories of environmental value (Figure 1.2).

An environmental asset's *total economic value* is the sum of its use value and non-use value. *Use value* refers to value derived from actual use of an environmental asset, while *non-use* or passive-use value refers to option and existence values. Use value, often broken down into consumptive and non-consumptive use value, is relatively straightforward. Examples of the former include drinking water, timber, wild game and recreational opportunities. Non-use values include many of the regulating and supportive ecosystem services. *Option value* arises from a willingness to pay for access to a particular environmental asset in the future and is based on uncertainties in future supplies, technologies and/or preferences. *Existence value* is the most controversial aspect of environmental valuation and refers to value placed on an environmental asset unrelated to any actual or potential use of the asset. Vicarious consumers of nature films or travel writing might ascribe existence value to particular ecosystems, as might societies who venerate specific natural sites for cultural or historic purposes.

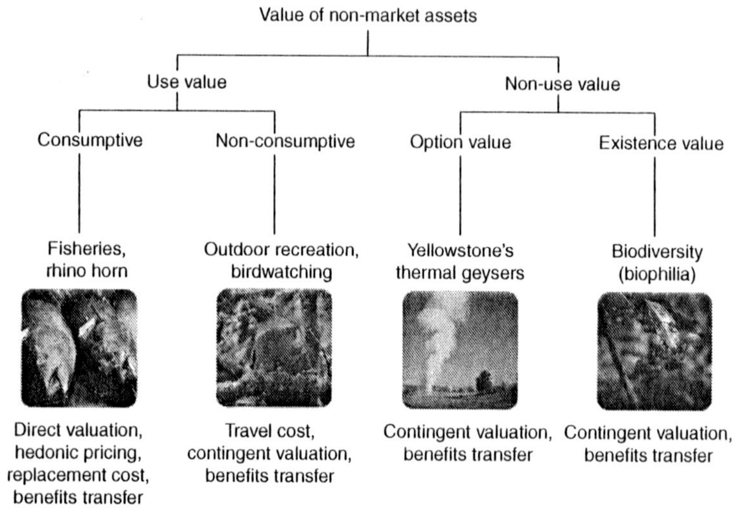

Figure 1.2 Value of non-market assets

A complete description of methods for valuing the environment is beyond the scope of this chapter. However, the following section provides an introduction to five of the most prominent techniques identified at the bottom of Figure 1.2. Before going further, however, it is well to note three things. First, market prices reflect only small (marginal) changes in demand. This isn't a problem when making decisions at the margin, but market prices can't really capture the impact of large changes (e.g. the collapse of marine food webs from overfishing or a catastrophic decline in wild pollinators brought about by climatic changes). Second, there will always be some ecosystem services for which there are no markets, and thus estimates from these techniques are likely lower bounds on values. (Moreover, because prices rise as a good or service becomes scarcer, market prices generally underestimate total values). Finally, note that one goal of valuing environmental services is to provide owners with incentives to conserve them. Providing the right incentives is not necessarily the same as valuing the services. We can provide incentives without valuing the services, and value services without providing incentives. Nonetheless, valuation does provide guidance in the development of proper incentives.

The hedonic pricing method relies on comparisons between the asset being valued and a set of comparable assets sharing similar characteristics. It is frequently used in residential housing appraisals, where a large set of comparable sales is available differing in certain key characteristics (i.e. distance from the beach, mountain views, etc.). By comparing two similar transactions, one with the amenity and one without, it is possible to tease out reasonable estimates of the values of certain environmental goods and services. Alternatively, one can compare productivities of enterprises in various locations. For example, a study of coffee farms in Costa Rica found coffee yields increased by 20 percent within a 1 km distance of forests, where wild pollinators (feral honeybees and stingless bees) nested (Ricketts et al., 2004). This method is relatively costly and time-intensive to implement, requiring large amounts of data to achieve robust statistical results. Also, markets must exist for the reference good or service (e.g. residential housing or coffee).

The travel cost method is often used to estimate the benefits of environmental recreational opportunities (e.g. hunting, fishing and boating), arguing that people's willingness to pay for the recreational opportunity can be estimated by the sum of the expenses needed to get there. Surveys are used to identify individuals' expenditures on travel, meals, equipment, lodging and the like, and a demand function is created from these results. A key assumption is that visitors would react to an entrance fee (if one existed) no differently than to an increase in travel expenses. Then, the total economic benefits obtained by visitors to the site can be weighed against other potential uses for the amenity (e.g. timber sales). Both the hedonic pricing and travel cost methods are 'revealed preference' or indirect valuation methods. They rely on expenditures related to, but not the actual environmental amenity. One challenge with the travel cost method is that it's difficult to take into account multi-purpose travel, i.e., visiting multiple sites on a single trip.

The *replacement cost method* is just what it sounds like – a method for estimating values based on replacement costs. A market-based technique, it seeks to identify how much it might cost to replace the environmental goods or service in question (e.g. $30,000 for a tree or $180 for a songbird). Although the replacement cost technique does not identify actual willingness-to-pay for these amenities, it does provide useful ballpark estimates of what they might be worth. The archetypal example of this technique is associated with New York City's decision to restore the Catskills watershed rather than purchase a costly water treatment plant. For decades, New Yorkers had enjoyed some of the cleanest water in the world from a range of foothills about 90 miles west of the city. In fact, the water was of such high quality – with a reputation of Evian or Perrier – that it was bottled and sold in the 1930s and 1940s. By the 1990s, however, development (sewage, runoff) and intensified agriculture (contributing fertilizer, pesticides and animal feces) had so degraded water quality that the US Environmental Protection Agency put the city on notice to fix the problem or build a new plant. In 1996, faced with a decision to build a new $6–8 billion filtration plant with $300 million in annual operating costs, the city cast about for alternatives. A year later, the city floated an environmental bond issue with the goal of restoring the Catskills watershed. The city used the proceeds to buy 100,000 acres of land to prevent development and control agricultural use, purchase conservation easements, and pay farmers not to grow crops or graze cattle along streams. All told, the city spent about $1 to $1.5 billion on restoring the watershed – 12 to 25 percent of the estimated cost to replace its valuable water purification services through technological means (Chichilnisky and Heal, 1998).

The contingent valuation (CV) method is the most controversial of the techniques for valuing the environment but also the only one that measures non-use values. The CV method uses surveys or experiments to discern what people are hypothetically willing to pay to acquire an environmental benefit or hypothetically willing to accept to bear an environmental cost. As such, it exemplifies the 'stated preferences' technique.

Most commonly, CV studies present participants with a binary choice between two different states of nature and are told that one of the alternatives can be obtained for a certain level of payment, or alternatively, that a certain level of compensation will be paid for the removal of an amenity. By varying the prices and using large samples, a demand function can be created and with it, the total economic value of the environmental asset in question.

When designing a study, good contingent valuation practitioners concern themselves with four key elements: defining the relevant population to be investigated (and constructing a robust sampling protocol), carefully specifying the environmental amenity to be valued and the method of payment (results are best when respondents are familiar with the asset being valued), collecting supplementary attitudinal and demographic data, allowing researchers to make sense of the results, and the analysis itself. Conducting a good CV survey of environmental benefits requires a substantial amount of development work, including pre-tests, focus groups and the like.

After completing a CV survey, it is helpful to compare the results with other valuation methods to determine the reasonableness of the estimate. For example, the author of this chapter once conducted a study using the CV method to solicit values for endangered species in Virginia (White, 1996). The results indicated Virginians were willing to pay an average of

$500 per person for the preservation of endangered bald eagles. Cross-checking these findings against other data, he found that it costs approximately $500 to raise a bald eagle chick from fledgling to adult at a nearby wildlife rehabilitation center, and that a bald eagle had a 'black market' value of around $2,500. The similarity of these three values corroborated the initial estimate arrived at using the CV technique.

The chief drawback of the CV method is its hypothetical nature; although it is market-based, because respondents do not actually purchase or give up the asset in question, disparities can arise between what they say they would do and their actual behavior. Well-designed surveys and experiments can help mitigate these differences, but the essence of the method relies on responses, not transactions. Other potential problems include strategic bias, in which respondents report extremely high or low values to exaggerate results, or aggregation bias, when participants ascribe the identical value to one or multiple ecosystem services. For these reasons, and because non-use or passive value is itself a controversial concept, CV studies are subject to greater scrutiny than the techniques described earlier. Nonetheless, they comprise the majority of studies included in the Environmental Valuation Reference Inventory (EVRI) described in the next section.

The benefits transfer technique uses existing valuation information for one good or service to estimate the value of a similar good or service. Although it is rarely the best choice for estimating environmental values, its use has grown in popularity in recent years due to the high cost of conducting original, site-specific research using the primary techniques described earlier. Analysts using this approach essentially identify economic estimates of environmental benefits at one location and then 'transfer' them to another site. The $33 trillion estimate of the annual value of Earth's ecosystem services mentioned earlier was obtained using this technique. The critical element in benefits transfer studies is to obtain *correspondence* between the study site and the policy or target site. Good correspondence occurs when the sites share similar biophysical characteristics and the representative human populations are expected to use or value the sites in similar ways. Oftentimes values at the study site are defined as unit values (e.g. per fishing day, per gallon of wastewater treated or per hectare), and then multiplied by the appropriate quantity at the policy site to obtain estimates of total value.

To facilitate and ease comparisons between benefits transfer analyses, a number of organizations have begun compiling databases of environmental valuation studies. The Environmental Valuation Reference Inventory (www.evri.ca), a joint project of Environment Canada and the US Environmental Protection Agency, is a good example of these sorts of resources and helps researchers to identify references offering greater correspondence and consistency.

Markets for ecosystem services

There are numerous options for the design and implementation of payment for ecosystem services (PES) systems. According to Wunder (2005), PES schemes share five basic characteristics. A PES scheme is:

1. A voluntary transaction where
2. A well-defined environmental service (or a land use likely to secure that service)
3. Is being 'bought' by at least one buyer
4. From a (minimum of one) environmental service provider
5. If, and only if, the environmental service provider secures environmental service provision (conditionality).

Payments for ecosystem services programs fall generally into three main categories: (1) public payment schemes, (2) self-organized deals and (3) open trading schemes.

Public payment schemes are often country-specific and include outright incentives to providers who adopt land management practices that preserve specific ecosystem services (e.g. no-till agriculture and riparian buffers), tax breaks to encourage certain behaviors (e.g. conservation easements) and payments made under philanthropic conservation programs (e.g. the 'debt for nature' swaps facilitated by some of the international environmental groups in the early 1990s). Government programs are the most widespread form of environmental service payments, and are likely to increase in the future as countries recognize the value of preserving critical natural assets and seek to identify ways of compensating those who provide them.

Self-organized deals are voluntary arrangements between private parties in which the beneficiary of the ecosystem service in question contracts directly with its provider(s). For example, Perrier Vittel, the French bottled water company, paid local farmers in the Rhine-Meuse watershed to adopt less intensive farming methods and improve animal waste management, in addition to purchasing 600 acres of sensitive habitat. Securing a supply of clean, uncontaminated water is obviously of critical importance to the company. The Nature Conservancy, an environmental organization, has taken this approach to the extreme, identifying areas of especially sensitive habitat around the world – and buying them. Self-organized transactions are motivated by a diversity of objectives both commercial and philanthropic.

Open trading schemes encompass both regulatory and voluntary markets. Regulatory markets are established through legislation and create demand by imposing a 'cap' on the amount of harm to which a particular environmental amenity may be exposed. For example, countries that have signed the Kyoto Protocol have agreed to limit their CO_2 emissions to specific levels. Power plants, cement plants and organizations regulated under the agreement must either reduce their emissions or purchase CO_2 allowances from other entities. This 'cap and trade' system effectively creates a market in carbon dioxide emissions, which, as we have seen, is an important element of global climate change. A robust market in CO_2 allowances exists in Europe, where most countries are Kyoto Protocol signatories and members of the European Trading System.

The United States is not a member of the Kyoto Protocol, but that has not stopped some US companies from participating in voluntary carbon markets. The Chicago Climate Exchange (CCX) was once the most visible player in this space, facilitating trades between corporations who voluntarily committed to achieving certain CO_2 reduction targets. Although the CCX abandoned its cap-and-trade program in December 2010, the Regional Greenhouse Gas Initiative (RGGI), a partnership among 10 Northeastern and Mid-Atlantic states that have agreed to limit CO_2 emissions, remains in operation, and California introduced a greenhouse gas cap-and-trade program in 2012.

Mitigation banking is another example of a regulated formal market for ecosystem services. Under the 1977 Clean Water Act, developers must mitigate the destruction or disturbance of wetlands, streams or endangered species habitat by preserving equivalent amenities at other sites. Wetlands mitigation banks have arisen to facilitate the establishment, certification and trading of endangered wetlands, and are, in fact, one of the more lucrative markets for ecosystem services. Mitigation banks are popular with developers due to their ease of use and with regulators because they often result in the creation of larger parcels of contiguous wetland areas, with positive environmental effects. More than 500 wetlands mitigation banks are now operating or in the planning stages in the United States.

Conclusion

Nature provides humanity with many resources critical for our well-being, but until recently, only a subset (e.g. food, fiber, timber, water) have been priced and traded in markets. We now recognize that supporting services – climate and disease regulation, water and waste treatment, soil formation, etc. – are declining and that one way of addressing the situation is to bring ecosystem services into our decision-making processes by assigning property rights and economic values. This chapter has briefly introduced the notion of ecosystem services, the methods economists use to value these services and various schemes for facilitating payments for these services. As population and consumption expand, and accessible natural resources continue to decline, the need for valuation and markets will only increase.

Forward-thinking businesses are already engaging with governments and non-governmental organizations to mitigate these risks. For example, beverage makers Coca Cola and SABMiller have partnered with the World Wildlife Fund to set standards, conserve freshwater resources and address critical water challenges. In response to the Securities and Exchange Commission's recent interpretative guidance regarding the disclosure of climate-change related risks, Goldman Sachs, GE and the World Resources Institute launched an initiative to measure water-related risks to companies and their investors. Companies participating in environmental markets may hope to gain preferential access to key environmental resources such as water, forests and waste treatment services. Opportunities also exist for companies to reduce costs and/or tap additional sources of capital. Firms may also see possibilities for new revenue streams (e.g. through the sale of carbon mitigation credits or wetlands banking).

Markets for ecosystem services are in their infancy, but will certainly grow in importance in coming decades. This is not to say that market prices can ever capture the entirety of nature's value to humans, or that prices equal value. Indeed, values obtained using the techniques described in this chapter should be considered *minimal* values for nature. However, valuing ecosystem services does bring humanity one step closer to recognizing the true cost of our consumptive activities, thus allowing us to make better choices regarding trade-offs between our various present and future desires.

Valuation of environmental amenities is of particular concern for citizens of developed economies such as the United States and Germany for two reasons. First, prices reflect scarcity. Without complete information regarding the relative scarcity of an asset (e.g. when assets remain unpriced) economic allocations are suboptimal and inefficient. Second, as natural assets become more scarce – as has been the case over the last

50 years – the wisdom of a growth economy is called into question. We should not seek to maximize wealth, but well-being.

Most Americans and Germans already enjoy standards of living well above the majority of Earth's people, and indeed, it does not seem plausible that all can ascend to such a level. Putting a price on the planet better aligns true costs with true benefits, increasing the probability that we might achieve a sustainable future.

Notes

1. Ecological economics is a relatively new discipline within the field of economics that recognizes the difficulties of allocating scarce resources on a finite planet. Ecological economists seek first to address problems of sustainable scale, then just distribution and finally economic efficiency. Ecological economists refer to conventional economics as 'empty earth' economics and their own discipline as 'full earth' economics.
2. Cf. *Steady-State Economics* (1991) and *Beyond Growth* (1996).

References

Chichilnisky, G. and Heal, G. (1998) 'Economic returns from the biosphere', *Nature*, 391, 629–630.

Costanza, R., d'Arge, R. and de Groot, R. et al. (1997) 'The value of the world's ecosystem services and natural capital', *Nature*, 387, 253–260.

Daly, H. (1991) *Steady State Economics*, 2nd ed. (Washington, DC: Island Press).

Daly, H. (1996) *Beyond Growth: The Economics of Sustainable Development* (Boston: Beacon Press).

Millennium Ecosystem Assessment (2005): Ecosystems and Human Well-Being: Synthesis (Washington, DC: Island Press).

Ricketts, T., G. C., Daily, P. R., Ehrlich, et al. (2004) 'Economic value of tropical forest to coffee production', *Proceedings of the National Academy of Sciences of the United States of America*, 101(34), 12579–12582, www.pnas.org/cgi/doi/10.1073/pnas.0405147101.

Vester, Frederic (1996) *Der Wert eines Vogels*, 2nd ed. (München: Kösel-Verlag).

White, M. A. (1996) 'Valuing unique natural resources: the case of endangered species', *The Appraisal Journal*, 64(3), 295–303.

Wunder, S. (2005) 'Payments for Environmental Services: Some Nuts and Bolts' (Jakarta, Indonesia: Center for International Forestry Research (CIFOR). Available at http://www.cifor.cgiar.org/publications/pdf_files/OccPapers/OP-42.pdf.

2

International Trade and the Environment: Does Globalization Create Havens of Pollution?

Peter Debaere

Introduction: The pollution haven hypothesis

The day was 30 November 1999. Thousands of protesters gathered in Seattle for a massive anti-globalization demonstration. The National Guard was called in to contain the masses and so was the riot police. The protest would turn out to be one of the most significant in the United States since the civil rights marches of the 1960s. The protesters formed a loose coalition of national and international nongovernmental organizations (NGOs) concerned with the environment and consumer protection, of labor unions such as the American labor movement represented by the American Federation of Labor and Congress of Industrial Organizations (AFL-CIO) and of religious and other groups. The protesters were marching toward the Washington State Convention and Trade Center where the third ministerial meeting of the World Trade Organization (WTO) was held. On the WTO's agenda was a new round of international trade negotiations that would include trade liberalizations in agriculture and services as well as questions of intellectual property rights protection. The anti-globalization protests quickly overshadowed the official negotiations and made clear the need for a broader public debate about globalization and its effects. The Seattle protests focused the world's attention on many questions that, at least in the United States, had been lingering since the North American Free Trade (NAFTA) negotiations of the late 1980s and early 1990s, which aimed to liberalize trade and investment between the United States, Canada and Mexico. It was high time to openly address key questions related to globalization such as: What are the gains from globalization, and are they equally distributed? Are there winners and losers? Are some of the international institutions overseeing the process of globalization such as the WTO, the International Monetary Fund (IMF) or the World Bank (WB) sufficiently democratic and transparent? Does globalization erode labor standards and, particularly relevant for this chapter, does globalization come at the cost of a clean environment?

The public debate triggered by what is sometimes referred to as the 'Battle of Seattle' was long overdue. Since the end of World War II, there has been a fairly steady and remarkable integration of the world economy that is often simply referred to as 'globalization'. Testament to this integration of the world economy are the steady growth of international trade in goods and services, the increasing capital flows between countries and the substantial international migration of people. To a large extent, these dramatic but gradual changes had not been the subject of many public questions before NAFTA and Seattle. In addition, some of the questions raised had not been the subject of much academic research either. In this chapter, I will analyze and discuss how globalization and the environment are intertwined. In particular, I wonder whether free trade is good for the environment and what it implies for advanced economies such as the United States and Germany.

The debate about the relationship between international trade and the environment is an important one. It is fueled by occasional reports about environmental degradation due to growing pollution in China, global warming, the destruction of mangroves due to rapidly growing shrimp aquaculture and exports, the relocation of polluting plants to countries with low environmental standards and so on. Environmentalists and the public at large wonder whether an open trading system may be inimical to preserving a clean, healthy and sustainable world. The debate about globalization is also relevant against the ever-stronger call, especially in European countries such as Germany, for stronger environmental regulations.

Discussing the link between international trade and the environment is not an easy task. There is plenty of room for misunderstanding, and a common language is often lacking. At the same time, there are no definite answers on many questions given that much of the research in the area is relatively new and often faces a paucity of data. The topic is also vast. Environmental quality has many dimensions – almost any human activity has environmental repercussions. To focus the discussion, I will therefore organize the chapter around what is often referred to as the 'pollution haven hypothesis', which perhaps most directly summarizes some major environmental concerns and expresses them in fairly general terms. The analysis of the pollution haven hypothesis also offers a useful framework to think about the recent discussions about CO_2

emissions, carbon taxes and carbon emission regulation in the context of climate change, which I will address at the end of the chapter.

In a nutshell, the pollution haven hypothesis states that globalization will give way to environmental dumping on a global scale. In a world with different national standards for environmental protection and increasingly open borders, the fear is that advanced economies such as the United States, Europe and Japan, which tend to be more sensitive to environmental issues, will off-shore their most environmentally damaging production to less-developed and emerging economies, which will become havens for pollution. In other words, in an integrated world economy, advanced economies would change the industrial composition of their production. They would produce fewer of the goods whose manufacturing process creates pollution themselves, and in a world of free trade in goods and services, the advanced economies would import the dirtiest goods from countries with less stringent environmental standards. In shifting their production toward cleaner industries and away from dirty ones, advanced economies would thus be able to enjoy the benefits of a cleaner environment at home, but it would be at the expense of other countries. It is feared that this off-shoring of pollution and the increase in international trade that comes with it could give way to an increase in worldwide pollution. Because of this, environmental concerns are often exhibit A in the anti-globalization movement.

In my discussion of the pollution haven hypothesis, I will draw on the tools of economics and work out consistent hypotheses and predictions based on economic theory. I will then confront these hypotheses with the data and the existing empirical literature.[1] The idea is to be as clear and precise as possible in the analysis and discussion. I will argue that trade liberalizations themselves are not likely to induce the type of industry composition that the pollution haven hypothesis predicts. If anything, trade liberalizations should lead to more production and exports of dirty goods by advanced economies (not by less-advanced countries with low pollution standards) because advanced economies have a comparative advantage (see later) in producing dirty goods. However, a legitimate concern remains. While trade liberalizations may not be the direct cause for increased imports of dirty goods from low-standard countries, tightening environmental regulations, especially in advanced economies, can trigger such imports in a free-trade environment. It should be stated, however, that the relocation of industries from countries with high environmental standards to those with lower standards need not be translated directly to movement from advanced to emerging economies, since there are also significant differences in pollution standards among advanced countries (e.g. the United States and Germany). Finally, when carefully discussing the empirical evidence, I point out that in spite of the public debate, studies for the United States indicate that technological progress – defined as the consistent reduction of emissions per unit of output – seems to have a far stronger impact on overall emissions and pollution reductions than any change in the industry composition associated with international trade. I close with a general discussion of the main implications of the analysis, and bring in the discussion of carbon taxes and global warming.

Before analyzing the particulars of the pollution haven hypothesis, let me frame the discussion by laying out some basic facts about globalization and in particular about international trade. This will help make clear why economists are typically reluctant to shut down globalization. It will also allow me to introduce some of the key tools for the analysis.

International trade liberalizations and globalization: Some facts

Since World War II, the world economy has become much more open. International trade between countries has fairly consistently grown faster than output, and countries consequently have traded ever-increasing fractions of their output with other countries. For reference, the ratio of exports to gross domestic product (GDP) was around 5–6 percent in 1950 for the world as a whole. In 2010, the IMF estimates that the ratio was about 30 percent. There has been a significant increase in the ratio especially since the 1990s.[2]

While there are many reasons for the integration of world markets, persistent trade liberalizations since the 1950s have played a non-negligible role. Tariffs or taxes on imported goods have come down dramatically worldwide. Estimates by the World Bank put the average tariff in manufacturing for high-income countries at 3.3 percent and at 11 percent for developing countries. Agricultural trade liberalizations have only been on the agenda very recently. This, in part, explains why they are much higher: 10.6 percent for the same high-income countries and 16.3 percent for developing countries.[3] Also non-tariff barriers such as quotas have been reduced. Instrumental in these trade liberalizations have been multilateral organizations such as the General Agreement on Tariffs and Trade (GATT) and the WTO, which replaced the GATT.

The GATT and now the WTO have impacted international trade in many different ways. For one, the GATT came into being to avoid the rampant protectionism of the interwar period that contributed to the Great Depression. The GATT has designed a set of rules and standards to guide international trade. Most prominent are the principles of the most favored nation (MFN) and of national treatment. Both principles are non-discrimination clauses. MFN states that member countries of the GATT/WTO should treat all their trading partners as their most favored nation and thus not practice discrimination. There are, however, important exceptions to the MFN principle. An ever-growing class of legal exceptions to MFN is formed by preferential trade agreements. There are currently more than 200 preferential trade agreements in force. In these, a restricted group of countries only reduces tariffs among themselves, but not necessarily with outside countries. The best-known preferential trade agreements are the North American Free Trade Agreement, the European Union (EU) and Mercosur.[4] Another category of exceptions is for developing countries that benefit under the general system of preferences (GSP). This system gives zero-tariff entry to markets in developed countries without requiring reciprocity. In part because of this non-reciprocity, tariffs in developing countries tend to be higher. The principle of national treatment, finally, urges countries to impose domestic and foreign goods to the same sets of rules within their borders.

A second way in which the GATT and the WTO have contributed to the growth of international trade is through the multiple trade negotiations rounds that they have initiated. These rounds have grown in scope and depth. They have brought even more countries to the negotiation table. Currently, the WTO counts virtually all countries of the world among its members. The trade negotiations also have covered ever-wider areas. Initially, negotiations were focused mainly on manufacturing. More recently, restrictions on trade in agriculture and services have been included in the negotiations. Because of this wider scope, negotiations have been increasingly cumbersome and have taken longer to complete. By way of example, the most recent Doha Round, which was launched after the 1999 Seattle protest mentioned at the beginning of the chapter, has still not come to a successful close as this chapter is written.

Now that we have an idea of the fairly dramatic increase in international trade since World War II, I want to briefly sketch the standard economic argument for free trade and against protectionism. It will make clear why economists are reluctant to restrict international trade too quickly, even in light of adverse environmental impacts.

The gains from trade: The economic arguments for free trade

Economists tend to favor free trade for a fairly straightforward reason. Free trade allows for the most efficient organization of production on a worldwide scale. Since such an efficient allocation increases total world production, it should, at least in principle, make all countries better off. As a starting point, let's study how gains from trade are generated.

Consider the key difference between a closed economy that does not allow for international trade in goods and services and an open economy that does. In a closed economy, sometimes called autarkic economy, whatever domestic consumers consume is produced domestically, and vice versa. In an open economy, on the other hand, a country's citizens can consume more of certain goods than can be produced domestically by importing from abroad. Similarly, consumers need not consume all goods that are produced within their borders because they can export to other countries. In a world with international trade, therefore, domestic consumption is decoupled from domestic production, which allows countries to specialize their production structure toward goods they can produce relatively well. This is of critical importance. Open economies can specialize: They will produce more of the goods they produce in a relatively cost-effective way and export those. At the same time, they will make less of the goods that are very expensive to produce domestically and import those instead. In this way, with free trade, world production will be organized more efficiently, and the same number of resources worldwide should generate more output, which is why, at least in principle, all participants should benefit from the bigger pie.

Tropical fruit is an obvious example of how international trade allows for a more efficient allocation of production compared to autarky. Consider the huge amount of resources it would cost a country such as Germany in a relatively temperate zone to grow tropical fruits such as bananas if its consumers wanted to eat them and international trade was not permitted. Contrast this with how easily bananas are grown in more tropical climates. Therefore, northern countries such as Germany should not try to produce tropical fruits. Rather, they should import bananas and in return export, for example, high-tech products that they produce in a relatively more cost-effective way than some tropical countries.

To make the argument about the economic gains even more tangible, consider the simple model and the graphical analysis of Figure 2.1. The graph depicts the production, consumption and international trade of one good, any good, between two countries. Without loss of generality, assume that the world at a given moment in time only consists of these two countries. To simplify the analysis of the gains from trade, I compare complete free trade and autarky, rather than a reduction of trade impediments such as tariffs. For now, we ignore issues related to economic

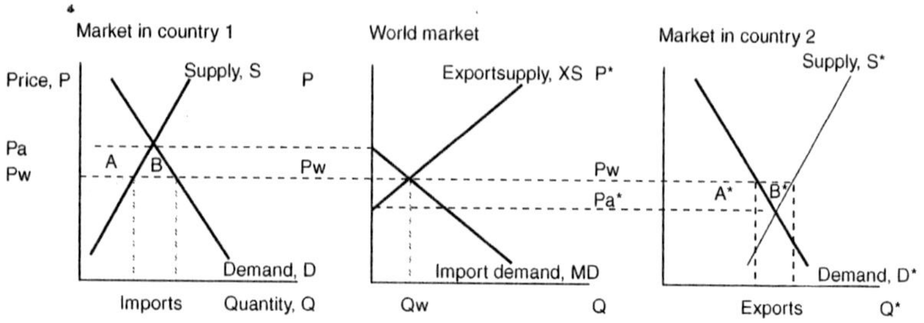

Figure 2.1 The gains from international trade

growth or technological progress. Once we get to the empirical discussion, I will bring those issues back in.

Consider the first pane, which represents the market for our good in country 1 with a domestic supply curve (labeled *S*) and a domestic demand curve (*D*). On the vertical axis, the good's price is measured (a higher point corresponds to a higher price); the horizontal axis traces the good's quantity (the farther to the right, the larger the quantity). As is common, the demand curve *D* is downward-sloping because it reflects consumers' willingness to pay: When prices are higher, consumers are willing to buy less of a good. As one can see, at a higher price (*Pa*), the quantity demanded is less than the quantity demanded at a lower price (*Pw*). The supply curve is typically upward-sloping. It reflects producers' costs. Producers supply more goods only when they can charge a higher price. As one can see in the left pane, the quantity that corresponds to the lower price *Pw* is smaller than the quantity supplied at a higher price *Pa*. In competitive markets, the equilibrium price that everyone ends up paying is determined by where demand meets supply. If the home country is not trading internationally, *Pa* is the equilibrium price. (Note that the subscript *a* refers to autarky.)

The third pane in Figure 2.1 depicts the market for our good in country 2 with a regular upward-sloping supply (*S**) and a downward-sloping demand curve (*D**). (The asterisk refers to country 2.) In the absence of international trade, domestic demand equals domestic supply and the equilibrium price amounts to the autarky price, *Pa**. Note that I chose country 2 as the country that has a comparative advantage in the production of the good: Its autarky price *Pa** is hence lower than that of country 1 (*Pa*).

Now, what happens when international trade is liberalized? We study the middle panel of Figure 2.1. It represents the supply and demand for our good on the international market. *XS* is the export supply curve.

It shows how much of a good country 2 is willing to export once the price is higher than its autarky price Pa^* – the vertical intercept of *XP* is Pa^* at which country 2 does not export at all. The *MD* import demand curve stands for how much country 1 wants to import from the country 2 once the price is lower than its own autarky price *Pa*. Needless to say, *MD* and *XP* are respectively derived from the first and second panel. The intersection of *MD* and *XS* determines the equilibrium amount *Qw* of country 1's imports as well as country 2's exports and the equilibrium world price *Pw* that prevails in both countries. (The subscript *w* marks the world equilibrium.) Not surprisingly, the world price is higher than the autarky price of country 2, Pa^*, which is why country 2 is willing to produce beyond what its domestic consumers would want. The world price, however, is lower than the autarky price of country 1, *Pa*, which is why consumers in country 1 are consuming more than under autarky.

In theory, both countries are better off under free trade than they are as closed economies. This can be shown graphically. Consider country 1. With free trade, consumers pay a lower price (*Pw*) and consume more goods. The surface of the trapezoid consisting of both areas *A* and *B* summarizes the gain for the consumers, which economists would call a gain in consumer surplus. Producers, on the other hand, lose out because they cannot all compete with the lower international price. Because of the stiffer competition from abroad, they end up producing less and they receive a lower price (*Pw*) for their product due to the competition. The trapezoid *A* summarizes the loss for the producers, or in economist terminology: the loss in producer surplus. Overall, however, the consumers' gains outweigh the producers' losses. Indeed, the area *B* therefore represents the gains from trade or the welfare gain associated with free trade for country 1. Now consider the situation in country 2. Here, the net benefits of consumers and producers are reversed, but overall, there is still a net welfare gain due to international trade. Consumers consume less of the good and also pay a higher price than under autarky, because the additional demand from country 1 has driven up the price. Compared to autarky, they lose the area A^*. However, the producers produce more than before and receive a higher price, which is why their total gain A^* and B^* exceeds the consumer loss by B^*. In sum, as this simple example shows, both countries are better off. Ultimately, both countries are better off because total world output is produced in a more efficient manner. Note that Figure 2.1 is only a partial equilibrium analysis because it only focuses on the trade in one good. Clearly, in order to pay for its imports, country 1 has to export another good that country 2 will import. In other words, a change in the industry composition in both countries will make the shift toward imports and exports possible.

Now that we understand the basic gains from trade argument, let's return to the pollution haven hypothesis and consider which country will end up exporting and producing more of the dirty goods that per unit of output causes most pollution, and which country will reduce its production of dirty products and increase imports.

Environmental concerns: Who exports the dirty good?

So far, we have not identified how trade liberalizations affect the production of the dirty versus the clean good. Ultimately, this is a very important issue at the heart of the analysis of pollution havens. If less advanced countries end up increasing their production of dirty goods after trade liberalizations, overall worldwide pollution should go up because pollution standards are less stringent in these countries. This is the nightmare scenario. Alternatively, if more advanced countries are the ones that end up exporting the dirty good under free trade, the prospects for world pollution seem better.

It is here we should ask what determines countries' comparative advantage and thus their ability to export certain goods. Mani et al. (1997) lists and ranks the dirtiest manufacturing industries in terms of air, water, metal and overall pollution. The industries that top the list tend to be very capital-intensive industries such as chemical, iron and steel, non-ferrous metals and so on. All these industries require lots of machines per worker. Because of their capital intensity, international trade theory predicts that advanced countries will be most able to produce dirty goods the cheapest, and hence advanced countries will be net exporters when trade is liberalized. Here is why. Compared to developing countries, advanced economies are more capital-abundant and labor-scarce. That is, per worker, advanced countries have much more capital (machines) than less-advanced countries. Because of this, wages are relatively high and capital relatively inexpensive in advanced countries, especially in the absence of trade. With a relatively high cost of labor and relatively low cost of capital, advanced economies therefore are fit to produce dirty goods cost-effectively because those types of goods require much capital and much less labor. Conversely, less-advanced countries are singularly well fit to produce labor-intensive goods whose manufacture requires lots of labor and little capital, which matches their low wages and their high price of capital.

Economic theory thus predicts exactly the opposite of the pollution haven hypothesis. In the wake of trade liberalizations, advanced countries (and not less-advanced countries) will be country 2 from Figure 2.1, which will export the dirty good and shift their industry composition toward producing the dirty good. Since advanced countries tend to have higher pollution standards, this is good news for world pollution but bad news for pollution in the advanced economies themselves. In the absence of technological improvements, advanced economies should face more pollution as they shift their industry composition toward dirtier goods.

Now what do the data tell us? There is a growing empirical literature that studies pollution and how it relates to international trade. This literature has focused mostly (but not exclusively) on advanced economies and has studied many measures of pollution. Particularly popular have been studies of air pollutants such as sulfur dioxide, nitrogen oxides, carbon monoxides and volatile organic compounds since data for these pollutants have been available for a relatively long period of time.

The data are consistent with our presumption that the more capital-abundant advanced countries should be responsible for most of the exports in dirty industries and that trade liberalizations *alone* should not give way to a relocation of the dirty industries to developing countries. Indeed, studies report that 80–90 percent of dirty products are produced by advanced countries. In addition, these observations are consistent with evidence that suggests that abatement costs that are incurred to minimize emissions are a relatively small fraction of the production costs and thus unlikely to change the prediction of international trade patterns, which are based on the relative abundance of capital versus labor. While the empirical findings do not at first seem to confirm the fears of the pollution haven hypothesis that trade liberalizations induce more pollution, closer analysis of the data and the hypothesis reveals there are nonetheless legitimate reasons for some concern in a globalized economy.

How do tougher environmental standards affect international trade?

Consider a world in which free trade prevails. What happens in such a world when environmental standards are tightened more in one country than in the other? We return to Figure 2.1. For simplicity, I assume that environmental standards are only tightened in country 2. Because of tighter environmental standards, producers in advanced country 2 face higher production costs. Needless to say, there are various ways in which higher standards can induce higher costs. For example, firms may have to buy scrubbers to reduce pollution in order to meet the tougher environmental standards; they may have to pay a tax on pollution or a tax for the waste they produce; in a cap-and-trade system, they may have to buy pollution permits that give them the right to emit certain levels of pollution. See for yourself what will happen. Because of the higher cost, the supply curve S^* in the right panel of Figure 2.1 will shift up and to the left – you can add $S^{*\prime}$ to the graph. This leftward shift indicates that to produce any given amount of a good, firms will now charge a higher price. Similarly, because of the higher production costs, the export supply curve XS will shift up and to the left – you can add XS' to the graph. The end result is that an asymmetric tightening of regulation tends to reduce exports from country 2. It also tends to alter the industry composition in both countries. As country 2 exports less, less-regulated countries will increase their production of dirty goods.

What does this analysis imply? When comparing more advanced economies with emerging economies, the former tend to be those with tighter environmental regulations. As we indicated before, with advanced economies having a comparative advantage in dirty industries, free trade in itself is unlikely to move the dirty industries off-shore to emerging economies. However, in an environment of free trade, tougher regulation and thus higher production cost in advanced countries will shift dirty good production toward lower-standard countries, confirming concerns by environmentalists. Note that the empirical literature supports this observation. Research by Levinson and Taylor (2008) shows, for example, that higher abatement costs in a country tend to increase the imports of dirty goods from abroad. As increases in standards are likely to be highest in advanced economies such as the United States, Japan and Germany, changing standards should increase pollution in emerging economies.

Note, however, that there are also significant differences in environmental standards and regulation among advanced countries. Comparing the United States, Japan and Germany, for example, it can be argued that Germany has taken the boldest actions since the 1990s toward a sustainable society: Germany has imposed eco taxes and has legislation on the books to reduce waste at the source and to induce recycling by manufacturers. Germany is actively promoting renewable energy and pushing global carbon emission reductions.[5] It should be clear that with diverging standards among advanced countries, it is an open question whether tightening of standards in one advanced economy moves dirtier goods to other advanced economies that have lower standards rather than to emerging economies.

In this context, it is worthwhile to briefly mention why countries sometimes choose to increase regulation and abatement costs, since it does matter for our discussion of the gains from trade and the benefits from a more efficient allocation of production with free trade. Pollution is sometimes called a negative externality, which means it is a by-product of production that negatively affects the welfare of the citizens of a country. As such, pollution is a social cost associated with the production of dirty goods, which is not included in the private production costs that firms face (expressed by the supply curve). By levying a pollution tax or by forcing abatement costs onto firms, governments sometimes try to have firms internalize the externality. As shown in Figure 2.1, higher production costs, for example, due to regulation or pollution taxes shift the supply curve S^* to the left. When the leftward shift in the supply curve S^* truly reflects the total cost of production (private plus social cost associated with pollution), the externality has been internalized. In this case, we could reapply our previous analysis of the gains from trade and find out how much producers benefit or lose from trade in the same way as we did before. With a supply curve that has shifted, the A^* and B^* areas would be a bit smaller than before due to the upward shift of S^*, but there would still be a net gain from trade.

Now consider the situation in the other, less-advanced country that does not enforce environmental regulation. In this case, there will be a significant gap between the social cost of production and the private costs that firms face: The supply curve will only reflect the private cost, excluding pollution. To properly assess the gains from trade, one would have to include the change in the social cost (pollution) associated with free trade. Since country 1 would end up producing less of the dirty good with free trade, its gains from international trade should be greater. Indeed, the less-advanced country gains as it can import the dirty good that it is not very efficient in producing from advanced countries. In addition, as it produces less of the dirty good, it will suffer less from local pollution

Technological improvements as deus ex machina?

Our findings so far indicate that advanced economies that have higher environmental standards are the likely net exporters of dirty products. However, under free trade, the asymmetric raising of standards, especially in advanced economies, is likely to change the industry composition in advanced economies. In fact, those countries may marginally reduce their net exports of dirty goods and consequently increase the production of dirty goods by low-standard, less-advanced countries. In what follows, I will place these findings in the broader discussion of technological progress in abatement to get a better sense of the magnitude of the potential impact on the environmental consequences of globalization. In this section, I follow in particular Levinson (2009), who focus on the United States.

For the United States and for other advanced economies, many pollution measures have been fairly consistently going down. Consider sulfur dioxide emissions in Figure 2.2. The lowest line (2) in the graph depicts the actual pollution of sulfur dioxide in the United States between 1987 and 2002 (pollution in 1987 is set equal to 100). As one can see, there is a significant decrease in pollution of 27 percent over the period.[6] How can we relate this reduction in emissions to the previous discussion and to the concern that a changing industry composition (in conjunction with increased imports) might increase pollution? It is here that we bring in the question of economic growth and technological change.

Many of the empirical analyses consider pollution (P) a function of the scale of production in a country (S), the technology that is used in the country (T) and the particular industry composition in a country (C) as in $P = f(S, T, C)$. Relatively popular is a rather insightful empirical decomposition of the change in pollution in recent years.

Researchers have decomposed the changes in overall pollution into three parts. They have looked at the (1) changes in pollution that can be

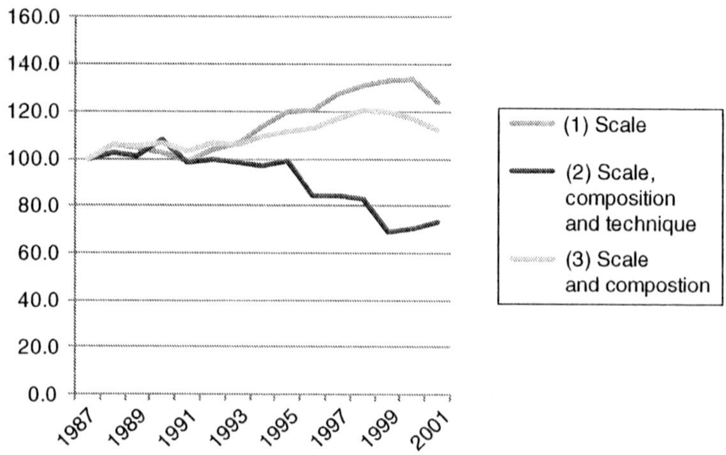

Figure 2.2 US sulfur dioxide decomposition since 1987
Source: Levinson (2009).

attributed to changes in scale or output (without any change in technology or any change in industry composition), (2) changes in pollution that are driven by changes in the industrial composition of a country's production (without considering any changes in scale and technology) and finally (3) changes in pollution that follow from technological change. While such decompositions do not give a causal explanation of the factors behind technological change, output growth or a changing industry composition, they do give a good idea of the relative contributions to overall actual pollution, of technological improvements, output growth and a changing industry composition.

Figure 2.2 shows this decomposition for the United States. As mentioned before, the lowest line (2) provides the actual pollution measured over the entire time period. It reveals a steady decline in sulfur dioxide and a decrease of 27 percent since 1987. The highest line (1) is a counterfactual. It shows that US pollution would have increased by 24 percent since 1987 in case the growth of US output had not been accompanied by a move toward cleaner technologies, and in the absence of a change in the industry composition of the United States toward cleaner industries. The key question at hand is to figure out what can 'explain' the drastic drop in pollution as exemplified by the difference between lines (1) and (2), and to figure out how much of this decline in pollution is due to better technology or to producing in cleaner industries. In particular, for the year 2001, pollution was 51 percentage points (24 + 27) lower than it would have been without cleaner technology (lowering emissions per unit of output) and without changing industry composition (a shift toward cleaner industries). Line (3) helps us disentangle the relative role of cleaner technology and cleaner industries in this dramatic reduction.

Line (3) shows how US pollution would have evolved if the shift toward cleaner industries had not been accompanied by the use of cleaner technology. As one can see, the shift toward cleaner industries while using 1987 technology would have increased pollution in the United States by only 12 percent. Consequently, the move toward cleaner industries in the United States closed a little less than half the gap between lines (1) and (2), or 23.5 percent to be precise (23.5 = (24 – 12)/(24 + 27)). This leaves 76.5 percent of the gap to be explained by cleaner technology, which is nothing but lower emissions per unit of output. In sum, the decomposition of sulfur dioxide, which is quite representative of other pollutants, indicated that technological improvements are the main engine through which emissions have decreased in the United States. Note that 'technological progress' is broadly defined in this context and includes in the particular case of sulfur dioxide for example, the use of low-sulfur coal (see Ellerman 2000).

In light of our discussion of the pollution haven hypothesis, we are particularly interested in the role of a change in industry composition toward cleaner industries, especially when this shift is associated with a shift toward importing more and dirtier products. We should note that the analysis does not explain the factors that drive the shift toward cleaner industries. We know, for example, that there has been a worldwide shift toward cleaner industries and that, next to trade liberalizations, differences in standards may also have played a role. The analysis, however, does give an upper bound: At the limit, 23.5 percent of the total pollution reduction is associated with a shift toward cleaner industries.[7]

Closing discussion: So, is free trade good for the environment?

As the pollution data for the United States and for many other developed countries show, there has been a steady decrease in pollution since the 1970s.[8] Technological progress (the use of cleaner technologies) has played a major role in bringing down pollution levels in spite of economic growth. This evolution is positive, to say the least. It suggests the possibility that cleaner technologies may have an important role to play in further decreasing pollution in less-advanced countries in the future as well. Recent empirical work for China already shows a decrease in emissions per unit of output (see Dean and Lovely, 2010).

The reduction in US pollution is, however, also associated with a shift toward cleaner industries. Interestingly enough, for the United States, about 23.5 percent of the reduction in pollution is associated with production in cleaner industries. About a fourth of this reduction has been associated with imports of dirtier goods that can displace domestic emissions. As noted, the empirical analyses do not reveal the exact reasons behind this increase in imports of dirtier goods. Trade liberalizations alone are unlikely to have been a driving force behind these reductions. If anything, trade liberalizations are likely to shift the production of the dirtier goods that tend to be fairly capital intensive toward cleaner, more capital-abundant advanced economies. However, tighter US regulations are potentially a reason why imported dirtier goods from abroad can displace domestic emissions of advanced economies in a world with relatively free trade imports.

Where does this leave us on the policy front? Economists are reluctant to use the usual trade policy tools such as tariffs or quotas to reduce the imports of dirty goods. The prime reason is that international trade and the specialization of production that it induces bring gains from trade. An additional reason why tariffs and quotas should be avoided is that specific domestic policy measures, such as domestic taxes and regulations that reduce pollution in the exporting countries themselves, create fewer distortions than international trade policy measures such as, say, tariffs, which try to enforce cleaner production from the outside.

The empirical evidence suggests that technological progress has a major role to play in pollution reduction. If anything, the available evidence calls for making technology transfers to other countries to facilitate the adoption of cleaner technologies abroad. In addition, the evidence begs for more research into the causes of technological improvements. There is, for example, the well-known Porter hypothesis. It posits that tougher regulations are instrumental in fostering technological innovation to such an extent even that countries with the toughest regulations will emerge as the main exporters of environmentally friendly products in the future because of the technological edge that regulation has induced. While this hypothesis can easily be interpreted as a call for government involvement and industrial policy, it is still waiting for broad-based empirical support. In light of the different approach that advanced countries such as Germany and the United States are taking with respect to environmental policy and regulation, it will be interesting to see whether indeed, the more pro-active German policies will make it the preferred location for environmentally friendly industries. Time and future research will tell.

... And what to do about carbon taxes and global warming?

Even though some would argue that the odorless and invisible CO_2 emissions that have been linked to global warming are not always considered pollution, the pollution haven hypothesis provides a useful framework to think through the discussion about CO_2 emissions and the suggested carbon taxes or carbon regulations to fight them. Like any pollution regulation or tax, carbon taxes and regulations will increase the cost of production in the country that imposes them. Just as in the case discussed above, advanced economies are likely to be the ones imposing carbon taxes before emerging economies do. The additional production cost in advanced economies will then give incentives for industries to relocate to countries with less regulation or without pollution taxes. The latter is often referred to as carbon leakage.[9] To the extent, however, that advanced economies have a comparative advantage in carbon-intensive industries, this effect will be mitigated. In addition, here also, a key role is expected to be played by technological progress to reduce emissions. There is, however, one important difference between CO_2 emissions and regular pollution. Air pollution, for example, is to a large extent local pollution where local communities directly suffer the consequence of the pollution in their local environment. As such local communities will have an incentive to address the problem locally and to internalize the externality with, for instance, a pollution tax. Global warming, on the other hand, is primarily a function of worldwide CO_2 emissions. A country's attempts to fight CO_2 emissions can therefore be undone if not all countries cooperate. In other words, a successful reduction of CO_2 emissions hinges upon international coordination between countries, which explains why a global reduction in CO_2 emissions is hard to attain.

Notes

1. A very good survey of the pollution haven hypothesis is found in Copeland and Taylor (2008); Chapter 13 in McLaren (2011), provides an accessible discussion of environmental issues for undergraduate econ majors.
2. Note that the export-to-GDP ratios for individual countries tend to vary with size: The export-to-GDP ratio of Germany, for example, is about double that of the United States, but about half that of Belgium. Note, however, that the export-to-GDP ratios for individual countries will all tend to reflect this upward trend of rising export-to-GDP ratios.
3. For a discussion, see Deardorff and Stern (2001); the data come from the WTO IDB database for 2005.
4. Regional Free Trade Agreement among Argentina, Brazil, Paraguay and Uruguay.
5. See Schreurs (2002) for an in-depth comparison of Germany, Japan and the United States.
6. For the other pollution measures, the graph is similar.
7. Levinson attributes to international trade (i.e. the United States importing more dirty goods) about a fourth of the pollution reduction that can be associated with to the change in industry composition toward cleaner industries.
8. Note that for many pollutants, longer time series sometimes reveal a steady buildup in earlier years when the economy was less developed before the recent decline in pollution. This hump-shaped curve has been called the Environmental Kuznetz curve.
9. Newell et al. (2013) find little evidence so far of carbon leakage, apart from a survey of managers from metals manufacturing and pulp/paper and cement and lime/glass industries who say they have moved their operations out of the area where compliance with European Emissions Trading (Europe's carbon market) is required. Note, however, that the research is only emerging and that the findings may well reflect the initial modest targets of Europe's carbon market.

References

Copeland, B. and S. Taylor (2004) 'Trade, growth and the environment', *Journal of Economic Literature*, 42(1), 7–71.

Dean, J. and M. Lovely (2010) 'Trade growth, production fragmentation, and China's environment', in R. Feenstra and S. Wei (eds), *China's Growing Role in World Trade* (Chicago: University of Chicago Press).

Deardorff, A. and R. Stern (2001) 'What you should know about globalization and the WTO', *Review of International Economics*, 10/3, 404–423.

Ellerman, A. D., P. L. Joskow, R. Schmalensee, Juan-Pablo Montero and E. M. Bailey (2000), *Markets for Clean Air: The U.S. Acid Rain Program* (Cambridge: Cambridge University Press).

Levinson, A. (2009) 'Technology, international trade and pollution from U.S. manufacturing', *American Economic Review*, 99(5), 2177–2192.

Levinson, A. and S. Taylor (2008) 'Unmasking the pollution haven effect', *International Economic Review*, 49(1), 223–254.

Mani, M. and D. Wheeler (1997) 'In search of pollution havens? Dirty industry migration in the world economy', *World Bank Working Paper* (16, April).

McLaren, J. (2012) *International Economics* (New York: Wiley).

Newell, R., W. Pizer and D. Raimi (2013) 'Carbon markets 15 years after Kyoto: Lessons learned, new challenges', *Journal of Economic Perspectives*, 27, 123–146.

Schreurs, M. (2002) *Environmental Politics in Japan, Germany and the United States* (Cambridge: Cambridge University Press).

3
New Directions in Green City Building

Timothy Beatley

The year 2008 witnessed a remarkable event: for the first time in the globe's history more humans lived in cities than that haven't. It is a truism now to describe planet Earth as the urban planet, but the realities of that fact have not sunken in and the fields of urban planning and design have not kept up with the special urban challenges we are facing today. There is much to do, many pressures and problems to face in navigating this global urban transition, and an ever greater need to re-imagine what cities are or could be, and to muster the creative new thinking and practice that is emerging. The trends are daunting: By 2070 some 70 percent of the world's population will live in cities. Innovation, resilience and sustainability can guide this new global urbanism, and there are many new ideas and emerging examples that will help show the way.

Global urbanization is a paradox: cities are engines for commerce and economic growth and development. As a recent UN Habitat report on the state of the world's cities says, 'No country has ever achieved sustained economic growth or rapid social development without urbanizing' (UN Habitat, 2008, p. x). With urbanization, however, comes increased wealth and material consumption, and an increased drawing down of global (and local) resources. But cities also represent the best hope for truly sustainable living: accommodating the needs of large populations through more ecological means of living: transit and walking, energy efficiencies from density, the possibilities of sharing many things in compact and urban settings. The urban engine that creates new commercial products and innovative ideas can focus this creative energy on solving the current problems. Any long-term strategy or plan for achieving global ecological balance, for living within the limits of the Earth, indeed reaching conditions where we are living more richly but restoring rather than degrading the Earth, will require concerted attention to cities and urban-built environments, which is the essential focus of this chapter.

Cities for people, not cars

Virtually every major advocate of sustainable cities imagines the future of cities as places where there is less reliance on and impact from private automobiles. The dependence of cars, especially in American cities and metropolitans areas, has many ecological, social and economic implications: many US cities remain in non-compliance for national ambient air quality standards largely because of auto-emissions and growth in automobility. Cars are extremely expensive on both private and collective levels (the average per-year cost of maintaining a car in the US is around $10,000, for instance), and are the major culprit in facilitating and encouraging urban sprawl and the pattern of highly land-consumptive and ecologically destructive land use and growth. Lower density urban form in turn leads to higher per-capita fuel consumption, and higher greenhouse gas emissions per capita, and makes more difficult investments in more sustainable public transport.

Sustainable and resilient cities of the 21st century and beyond will be those that invest in a diverse and robust network of alternatives to the private automobile, including public transit, walking and bicycles, among others. Public transit makes sense on many levels and is the least energy-intensive and environmentally destructive method for moving large numbers of people around in cities. There are increasingly many creative methods being employed in cities around the world to invest in transit that makes sense for specific settings and context. For cities in the developing world, in Asia and Latin America, use of bus rapid transit (BRT), for instance, has proven to be an efficient and highly cost-effective form of transport, compared with the heavy-rail and underground metro systems that many first world cities have constructed. The systems rely on bus-only corridors, with metro-style stations (with pre-payment), providing quick mobility, and contributing to the solution of air quality problems in many large cities. There are now a number of Asian and Latin American cities that are successfully using (and expanding) BRT systems, including Bogota (TransMilenio), Mexico City and Jakarta, among others. In February 2011, Mexico City opened its third BRT line, expected to provide service to another 120,000 passengers (pushing total ridership over 600,000 per day). The total system now includes 67 kilometers of busways and 113 stations (Embarq, 2011).

Unique urban conditions can (and should) in turn lead to new and different forms of transit. New elevated gondola projects have been completed, for instance, providing new mobility options for residents of favelas (slums), and the very poor, in cities such as Caracas, Medillin and most recently Rio de Janeiro.

Sustainable cities will increasingly need to be understood as cities where people can experience place and community, by walking and getting out of their cars. This is captured to considerable extent by the work of Jan Gehl, the Danish guru of walkable cities, whose new book is aptly entitled *Cities for People* (Gehl, 2010). Creating the conditions for pedestrian cities means compactness and density, and mixing-use, multifunctional forms of urban development. Northern European cities, such as Stockholm and Oslo, have emphasized patterns of dense contiguous growth, and German cities, such as Freiburg, have a long and exemplary history of steering development into dense growth areas along the city's tram lines. In Freiburg, the experience of walking is reinforced through investments in unique place elements, notably that city's network of water channels, which new buildings and development must extend, and which contributes much to the delightful experience of walking in that city.

While there has been a growing consensus among urbanists and environmentalists alike about the importance of compactness and density (e.g., see Owen, 2010), the extent of the density and form it takes remains an open debate. As urban land becomes scarcer, many argue that more vertical urbanism, exemplified more by Asian cities, such as Hong Kong and Singapore, will be necessary. North American examples, such as Vancouver, have shown well that vertical growth (density accommodated through high rise towers) need not undermine quality of life but can enhance it, and can occur at the same time as investments in urban amenities of various kinds, including streetscapes and pedestrian amenities. Vancouver's creative approach to urban design, wrapping vertical towers with low-rise buildings, wide sidewalks, double rows of street trees and a highly interesting, mixed-use urban environment, as well as investing in public transit, shows what is possible.

New concepts for reconfiguring and redesigning urban streets abound, with significant progress from pilots to mainstreaming many of these new ideas. One of the more interesting ideas is the notion of naked streets and naked intersections, or sometimes referred to under the moniker of shared space. Thinking and practice here developed in parallel in the Netherlands and Australia, and holds basically that efforts should be made to take away signage, lane stripping, lighting and other guidance and instructions to cars. To most Americans, this seems counterintuitive, if not downright dangerous, but these efforts do seem to slow traffic and do seem to positively adjust streets in favor of pedestrians.

One of the more notable examples of a naked intersection can be found in the Danish city of Christiansfeld. Several years ago I made a pilgrimage to this city to witness first hand this example, and to film the intersection, which had been converted from a conventionally lighted four-way intersection to one with no lights and virtually no signage. The results are startling. I watched (and filmed) for several hours as cars, fast and slow, did a delicate dance as they moved through this intersection. Drivers made eye contact, gestured and otherwise navigated safely. No accidents occurred on that day and in fact the evidence from this intersection, as with most shared space retrofits, is that the accident rate goes down. Of course, partial explanation for success here have to do with the traffic rules (especially that traffic from the right has the right of way) and that the brick and stone places at the intersection sends its own form of signal (slow down, this is public space). This highlights that there are many different urban interventions that can help to humanize and enliven urban streets and spaces. Paving and bricking investments that help to create the feeling that urban spaces are outdoor living rooms is one idea, and investments in public art can do much to make walking in urban environments interesting. Melbourne has taken many of these steps, with dramatic results, creating new public spaces (such as Federation Square), installing bluestone pavers, new street furniture and investing creatively in public art (including a laneways art initiative to entice pedestrians to visit alleys and laneways they would otherwise miss or avoid). The result is what Jan Gehl has called the 'Melbourne Miracle'. with dramatic increases in the numbers of people strolling, sitting, eating at outdoor cafes (the city went from only two outdoor cafes in 1973 to 356 in 2004).

More recently a number of US cities have sought to pedestrianize downtown spaces, and to make it easier to create temporary parks and green spaces out of parking and auto spaces. Notably New York City recently decided to keep closed to car traffic, portions of Broadway that had been closed temporarily for café and pedestrian use.

In the UK, the shared space movement has led to the designation and funding of a number of so-called Home Zones, where significant traffic calming and street interventions and re-designs have occurred. Even more recently, a non-profit called *Sustrans* has created and is piloting an approach it calls DIY Streets. With goals similar to Homes Zones, it is intended to create shared spaces and more pedestrian-friendly streets and community spaces but at a much lower cost, and through a more grassroots, neighborhood-based model, where much relies on recycled materials and volunteer labor from the neighborhood (see Sustrans, undated).

There are many other creative ways by which cities are seeking to discourage car ownership and use, recognizing the economic, environmental and other negative consequences of car-dependence. Shifting the economic signals is an important step. In many European countries, car ownership and use is has been significantly discouraged through taxes. In Denmark, new car purchases are subject to a 180 percent vehicle registration tax, a quite significant financial disincentive to purchasing a new car. [Electric vehicles, however, have been exempted from this tax, and recently the Danish government extended the exemption through 2015 (Copenhagen Post, 2009).] As a result, Copenhagen has one of the lowest rates of car ownership among European cities, with an estimated 208 cars per 1,000 population (Travel and Transport Research Ltd, 2005). Table 3.1 compares Copenhagen's rate of car ownership with other selected European cities, and shows the relatively successful efforts of this city, a result of the combination of providing walkable urban living conditions and investing in excellent alternatives to the car (transit, including new metro lines, bicycles). Few would argue that the quality of life is lower in Copenhagen because of the reduced dependence on cars – just the opposite is certainly true.

Copenhagen has emerged in the last two decades especially as an innovative leader in promoting bicycles as a viable mobility option. The city especially has shown a commitment to investing in the infrastructure of bicycles – lanes and movement spaces for bicycles, for instance, that makes it safe and easy to get around the city by bicycle. Most intersections in the city provide separate lighting for bicycles, and provide a nine-second advance green light over cars – that is, bicycles get to move through most intersections significantly ahead of car traffic. In 2007, the city even synchronized the intersection lights along a 2.5-kilometer stretch of the busy Nørrebrogade, a major bicycle corridor, further

Table 3.1 Car ownership per 1,000 population

Rome	665
Stuttgart	523
Madrid	431
London	365
Barcelona	350
Copenhagen	208

Source: Travel and Transport Research (2005).

speeding along bicycle commuters. Copenhagen has also recently initiated its so-called Green Cycle Routes initiative, creating new bicycle commuting routes through or alongside parks and green areas.

What is remarkable in the last five years, especially, is the tremendous progress made by many cities that would not be immediately identified as bicycle-committed. These include North American cities such as Montreal and Denver, and European cities such as Paris and London. In London, historically not a very bike-friendly city, major new commitments have been made, including a plan to create 12 Cycle Superhighways, bicycle commuting corridors that connect outer boroughs with central London. Supported through a partnership with the Barclays bank (they are called Barclays Cycle Highways in fact), the first two ... were unveiled in the summer of 2010, and another two launched in summer of 2011 (Transport for London, undated).

And bicycle mobility offers an affordable option in many cities in the developing world, though a struggle exists in such cities to overcome the bias in favor of automobility and the need to accommodate increases in cars and roads in order to ensure economic progress and development (and an equally strong bias against bikes as a poor and backward form for mobility). Efforts to curtail bicycle rickshaws in cities like Dhaka, Bangladesh, reflect this value structure, despite the fact that in that city there are nearly half a million rickshaws in use, providing a very important mode for moving vast amounts of people and materials and goods around the city, and one that provides economic livelihoods for as many as a million residents. Good examples of accommodation of bicycles can be seen in some Latin American cities such as Bogota and recently Mexico City, the former famous for its *Ciclovia*, where each Sunday it closes much of the city to car traffic, giving city streets over to bicyclists and pedestrians. While estimates vary, by one count, more than two million residents of Bogota are outside enjoying this event (with the assistance of a network of paid bikewatch *guardians*; Power, 2010). Cities, then, face serious choices not just about mobility, but more profoundly the quality of life they wish for their citizens.

Biophilic and green cities

Green cities can also literally be green: design to preserve and restore a variety of ecological and natural features, from forests and wetlands to river and riparian systems. There is growing recognition that modern cities, even very dense cities, can and must also provide daily access to nature, and that urban exposure to nature is essential for a productive, happy,

Table 3.2 Dimensions of biophilic cities (and some possible indicators thereof)

Biophilic conditions and infrastructure

- Percentage of population within a few hundred feet or meters of a park or green space
- Percentage of city land area covered by trees or other vegetation
- Number of green design features (e.g. green rooftops, green walls, rain gardens)
- Extent of natural images, shapes, forms employed in architecture and seen in the city
- Extent of flora and fauna (e.g. species) found within the city

Biophilic behaviors, patterns, practices, lifestyles

- Average portion of the day spent outside
- Visitation rates for city parks
- Percent of trips made by walking
- Extent of membership and participation in local nature clubs and organizations

Biophilic attitudes and knowledge

- Percent of residents who express care and concern for nature
- Percent of residents who can identify common species of flora and fauna

Biophilic institutions and governance

- Priority given to nature conservation by local government; percent of municipal budget dedicated to biophilic programs
- Existence of design and planning regulations that promote biophilic conditions (e.g. mandatory green rooftop requirement, bird-friendly building design guidelines)
- Presence and importance of institutions, from aquaria to natural history museums, that promote education and awareness of nature
- Number/extent of educational programs in local schools aimed at teaching about nature
- Number of nature organizations and clubs of various sorts in the city, from advocacy to social groups

Source: Beatley (2010a).

healthy life. Building on E.O. Wilson's notion of *biophilia* – that we have coevolved with the natural world, and need this connection and affiliation with nature, there is growing support for the idea of *biophilic cities* (see Beatley, 2010b). While a more complete definition of, and a set of metrics for measuring a biophilic city are contained in Table 3.2, the essential idea is that of cities that create the conditions for contact with nature, and for nudging urbanites to spend more time in close contact with outdoors. As Table 3.2 suggests, it is partly about urban form and planning, but also about programs and enticements, governance structures and much more.

Nevertheless, the physical planning is a key component. Creating walkable, pedestrian cites, as described above, will take us much of the way, as investments in mixed-use, transit-committed cities will translate to cities and living environments where residents spend more time outside and walking and less time in cars. But there is more needed still. There are important biophilic design ideas at every scale, important things that can be done from rooftop to region. Land use and urban form are important, and some of the best examples of biophilic cities can be found, for instance, in Scandinavia, where a combination of compact urban form with the securing of large blocks of natural land creates conditions in which urbanites are in very close proximity to nature. Copenhagen's famous Green Fingers regional plan, for instance, steers growth to compact towns and urban centers along the spines of it rail system, at the same time setting aside large green wedges between them. Helsinki has a comprehensive and connected green network, allowing one to travel from dense urban neighborhoods in the center of the city to old-growth forests on the periphery. A biophilic city ideally provides a hierarchy of parks and natural areas, from green courtyards and small pocket parks, to large expanses of woodlands and nature.

Oslo is one of the most biophilic cities in the world. It has a long history of guiding growth into its center city, leaving some two-thirds of the city in protected forest, or what in Finnish is referred to as *marka*. And the forest is an important destination for residents, an important recreational resource, with many residents visiting it and spending time there, even in winter. A new green plan for the city lays out an even bolder vision for the future. Most ambitiously, the city's has adopted the goal of *daylighting*, or bringing back to the surface, the eight streams that run through the city, connecting the marka to the north, with the fjord to the south. Already one major waterway, the *Akerselva*, has been restored, with beautiful walking trails along side and around the river, at several points with spectacular waterfalls, a remnant of the former industrial uses of the river.[1]

Biophilic cities are cities that also understand, celebrate and utilize the unique climatic conditions there. European cities have been much more effective at incorporating regional climate and weather as a key design factor in their spatial planning. Notably German and Austrian cities include significant climate sections in their spatial plans. In Freiburg, for instance, importance has been placed on preserving the replenishing airflows and breezes traveling through the city, down from the Black Forest. In certain parts of the city, buildings must be designed and configured to permit the continued flow of these breezes. Projects such as the new growth area, Reiselfeld, have been designed so that building massing and orientation permit flow-through for these cooling winds, and streets have been oriented east to west to facilitate flow. As most critical air movement has been found to be above ground, buildings in key areas in Freiburg are not permitted to be any higher than 12.6 meters (Daseking, 2011). Other cities, such as Stuttgart, have included similar provisions, in this case a regional effort focused on protecting the cool air movement zones, or 'ventilation corridors' or 'lanes' in this metropolitan region. In both Stuttgart and Freiburg, protection of surrounding forests and green spaces further help to cool urban areas, and efforts at protection of trees within the urbanized areas are also a key (e.g. Sustainable Cities, undated).

There has been much new and creative work to integrate green elements into buildings and urban neighborhoods. Green rooftops are now common features, even in North American cities such as Chicago (more than 450 built or in process) and Toronto (which has now adopted a mandatory green rooftop requirement for certain types of roof). Green walls and vertical gardens, such as those designed by French botanist Patrick Blanc, provide many similar benefits, cooling buildings, retaining storm water, providing new habitat and inserting a degree of beauty (see Beatley, 2010b).

Cities like Seattle have been pioneering new and innovative approaches to storm water management, using techniques of so-called low impact development (LID), which eschews large-scale engineered urban storm water retention and collection systems in favor of more decentralized, natural approaches that retain (and celebrate) storm water onsite, through rain gardens and bioswales, green rooftops and planting of native shrubs and vegetation.

Seattle also now implements a unique Green Factor requirement, stipulating that new commercial development and multi-family housing in certain designated (Green Factor) zones in the city must meet a minimum 0.5 on a green feature scoring system. Projects achieve points through including features such as green rooftops, vegetated walls, tree canopy coverage and permeable paving. Special bonus points are available for use of drought-tolerant and native plants, and for 'landscaping in food cultivation' (See City of Seattle, undated).

Imagining biophilic cities is also about fostering a sense of wonder and an active sense of being a member of a larger biological home, a larger community of life. Connecting to the nature around us, understanding how urban homes and buildings are situated in watersheds and bioregions, recognizing and feeling a closeness with the native flora and fauna, will also be important challenges for city planners in the years ahead. Part of the agenda of biophilic cities, then, is to understand cities as multi-sensory environments, where the feel of wind and weather are to be appreciated, where birds and trees and flowers are recognized, where the sounds of the evening are acknowledged and appreciated and where urbanites don't feel detached or disconnected but very much a part of the living city in which they reside. Larger cities often make this difficult, of course, for instance, massive amounts of urban lighting make it difficult to see the night sky, and globally an increasing percentage of the world can no longer see the Milky Way. Following the Northridge Earthquake that struck the Los Angeles region in 1994, for instance, police departments received numerous calls from perplexed residents wondering about the 'strange sky' – simply the Milky Way that was visible as a result of the power outage, showing surprised urbanites who either could not or chose not to look around at the wonder above.

How the rapid urbanization and rapid shift to cities that is occurring in much the world can also be respectful and even restorative of nature will be an important goal and challenge. And how can urbanites be challenged to see themselves as ecological citizens of a place, caring about its condition and working on its behalf?

Urban biophilia also suggests the value and importance of planning and designing cities that learn from and mimic nature and natural systems. Here there is an extensive and growing body of research on biomimicry, learning, as Janine Benyus has so eloquently said, from the 3.8 billion years of research and development that nature is. There are now countless ways in which we have designed objects and products, and increasingly buildings and cities, based on the insights of nature.

Green architect Bill McDonough is famous for issuing the design plea that every building should function *like a tree* and by extension every city as a forest (see McDonough and Braungart, 2002). There is much new work on building technologies and materials so that the concept of a truly living building is increasingly feasible. New paints for exteriors have been developed so that they actually serve to clean urban environments, not unlike the pollution cleansing and oxygen-producing values of trees, for instance.

There are now some dramatic examples of green urban buildings modeled after nature, for instance, the Eastlake Centre, in Harare, Zimbabwe, where the building's ventilation system is inspired by and modeled after African termite nests. A new office building is under construction in the Chinese city of Tianjin: the Sinosteel International

Plaza, is designed with a honeycombed exterior, utilizing the strength of this hexagonal shape in nature and reducing requirements for interior structural supports (Diaz, 2010; see MAD architects).

More recently, a company based in Brooklyn, New York, has developed a product called 'Solar Ivy', a system of façade-mounted photovoltaics, producing power from the sun, 'that mimics the form of ivy and its relationship to the environment' (SMIT, 2010, p. 1). 'Flexible photovoltaic "leaves" flutter and shift in the wind while converting solar energy into electricity that feeds into a grid tie inverter or charges batteries for off-grid storage.' The second generation product is even designed to harvest the energy produced by the fluttering of the ivy in the wind. And the company producing them, Sustainably Minded Interactive Technology (SMIT), makes the biomimicry connection clear: 'Solar Ivy is an integrative solar energy solution stemming from the resilience and structural ingenuity of plant life found in the natural world' (SMIT, 2010, p. 1). If only the solar ivy could grow and reproduce itself in the same way as natural ivy, increasing its solar production along the way, then the renewable energy value would be incalculable.

Edible cities

Food has emerged in the last few years in a big way on the urban planning agenda in northern industrialized cities, in part a response to increasingly popular and consumer support for organic and local food. There has been a marked rise in city markets and farmers markets, and other creative strategies for re-connecting farmers and urban consumers (e.g. community-supported agriculture, community-supported fisheries, metropolitan buying clubs). There has been a tremendous desire to reintroduce food growing and livestock to cities, and many American cities having recently modified their zoning codes to permit these uses (e.g. major examples include New York, Toronto, Chicago and most recently San Francisco). Arguably, agriculture never disappeared from Asian cities, in the same way it did in American and European cities. That said, it can be argued that this renewed interest in food, and push to plan for and accommodate urban agriculture in urban environments is nothing short of a radical new way of thinking about modern cities: cities are not simply black holes of consumption, but can be re-conceptualized as bountiful places, where at least some (if not all) of the food needs of its residents are satisfied from very local production: peri-urban farm fields, but also production in interior community gardens, urban greenhouses and rooftop gardens.

In the US context, cities that have been shrinking in population have offered unusual opportunities to advance this urban agricultural agenda. Detroit represents an extreme example, with a present population of less than half of what it had at its peak in the 1950s. About one-third of the city, about 40 square miles, is now vacant, offering unusual opportunities to re-imagine the city. An initiative of present mayor Dave Bing, called Detroit Works, is working to identify and densify five to seven population centers, while curtailing public investments and services outside these areas. Already nature has returned to many of these emptying neighborhoods, and in many east side neighborhoods, larger tracts of open land are emerging as residents leave and houses gradually burn or fall. What has resulted in many parts of Detroit is what Gallagher refers to as urban prairies. More by accident than design, populations of ring-necked pheasant have returned, and recently beaver have been sighted for the first time in decades.

John Gallagher in his 2010 book *Reimagining Detroit*, argues that the city has unusual opportunities to actively restore and heal its landscape. He proposes daylighting streams (especially Bloody Run Creek), planting more trees and reintroducing wildlife into the city. Leaving more land in the city in a natural condition has many advantages, including enhancing quality of life for those remaining in the city, providing significant climate and cooling benefits and saving the city money. As Gallagher concludes: 'A city can benefit economically by returning more land to nature even as it builds up other parcels nearby. A city can shrink and grow at the same time' (Gallagher, 2010, p. 117). The city of Detroit, then, has the potential to emerge as a model for how to gracefully and thoughtfully shrink a city, demonstrating a palette of creative city design ideas for the 21st century post-industrial (post-carbon) city. Gallagher believes part of where to begin is with attitude – that smaller need not mean lesser, but rather the basis for a more sustainable, innovative city, and one with a higher quality of life.

One thing is certain, then: agriculture will be an essential part of any future vision of cities. And the evidence for creative work is all around, and in many of the other cities facing similar though less severe economic conditions. In Milwaukee, Will Allen's vision for urban food production can be seen in a new aquaponics operation, Sweet Water Organics, producing Perch and Talapia in former warehouse spaces. In New York City, the 40,000 square foot rooftop of the Standard Motor Products Building is home to an organic urban farm run by the company Brooklyn Grange. Also in New York, two young designers have invented a kit-based system for growing food hydroponically in windows, called the Windowfarms Project, and already some 17,000 individuals are using them (see www.windowfarms.org). In Cleveland, a large scale (5½ acres) hydroponic greenhouse is in the works, organized as a worker cooperative and expected to produce by one estimate 5–6 million head of lettuce and perhaps 300,000 pounds of herbs.[2] Cities as diverse as London and Oakland have developed comprehensive community food strategies, and whatever the economic health and population dynamics there, these very urban environments will increasingly be understood as opportunities for growing food.

Cities with sustainable urban metabolisms

As the global flow of food demonstrates, urban population requires many kinds of resources and goods, and these inputs, as well as the outputs at the other end (in the form of municipal solid waste, air and water pollution, carbon emissions) can be seen as a complex metabolism. Analogous to the functioning of a human body, cities and regions have living metabolisms as well, requiring certain inputs to survive, generating wastes on the other end. Increasingly urban sustainability envisions cities that shift from the usual bloated metabolism, linear and dependent on long-distance supply lines, to something quite different. A sustainable metabolism is first one that seeks to reduce the size of the flows or the throughput. Efforts at reducing energy consumption, more efficient lighting systems and incorporating more extensive natural lights in buildings are examples of steps that will help. Second is the need to shorten supply lines, sourcing and producing more of what cities need closer by, where there is greater transparency about how and by whom these goods and materials are provided. The local food, urban farming trends discussed above are examples of this, but there are many other ways in which urban metabolism can be more localized and regionalized, and virtually every product or service offers the potential to be sourced or produced locally (or regionally).

The economic benefits are considerable: Jane Jacobs talks about the need for what she called 'import substitution': the fewer dollars sent out of the community for things that could be produced locally (or regionally), the more support for local suppliers and producers and the greater the multiplier effect of these dollars (Jacobs, 1970).

One particularly promising sector in many US urban regions is wood. As we grapple with the global demand for wood, tropical forests and its spur to tropical deforestation, new and creative businesses and business models are emerging in a number of cities to support at least some of this wood from local sources and in more sustainable ways.

Solar cities: Post-carbon cities

There is now an unusual amount of activity around designing, planning and re-imagining what cities will look like an how they will function in a post-carbon world. Many of the key features have already been mentioned, including investing in walkable transit cities and reducing reliance on autos, as well as producing food and materials more locally.

But more will be required and cities will need to lead the way in shifting away from dependence on fossil fuel energy. Indeed, there is considerable new thinking about cities, imagining that they can be net producers of energy and that all buildings and elements of the urban built environment can be re-imagined as energy producers. The ideas of social theorist and futurist Jeremy Rifkin are especially compelling (see Rifkin, 2010). He speaks in terms of a suite of actions (pillars) that must occur together to shift quickly into a new energy future. At the heart of this is a shift to a *distributed* network of renewable energy, reimagining every office building and home as a mini power station, and in turn igniting what Rifkin calls the democratization of energy. He argues against focusing on large centralized renewable production approaches (sun energy from Greece, wind from Ireland), and the need to transmit this energy through long distances, though he recognizes their value in the short term as a bridging approach. He believes there is renewable energy everywhere in the world, wherever we live: solar, wind, geothermal, ocean tides, small hydro and even garbage.

Rifkin tells me there are 191 million buildings in the European Union and they could all be producing power. The trick is to find ways to store and share this energy. The first is accomplished through a variety of potential storage technologies, though hydrogen Rifkin believes holds the greatest promise. Plug-in vehicles will also help with storage. Sharing energy will be accomplished through highly interconnected smart grids (what Rifkin calls an 'intergrid') that functions like the Internet, decentralized and collaborative. In this vision, 'millions of buildings are collecting energy, even a little bit of surplus, ten per cent, twenty per cent ... even two per cent that they don't need at 8 o'clock in the morning. They store it and ship it. Grid IT allows us to have this distributed power that exceeds anything you could ever imagine with nuclear, coal or natural gas plants' (Rifkin, 2010).

The vision of positive energy buildings, buildings as power plants, that produce more power than they need, is no longer a pipedream but a concept that is being put into practice in many places. Rifkin often cites the example of the GM factory, in Aragon Spain, that produces energy sufficient for some 4,600 homes from its rooftop array of photovoltaic panels. The Elithis Tower, in Dijon, France, is purported to be the world's first positive energy office building. The 54,000 square foot building, completed in 2009, was designed from the beginning to use a miniscule amount of energy compared with conventional office buildings (only about 20 KWh per square meter, about one-twentieth the French average) and then produce more energy than it needs to operate, thus achieving a 'positive' energy balance. The design is interesting, with a distinctive 'solar shield' covering the building's south exterior, allowing in extensive daylight but providing shade and reducing heat loads in the summer. The relatively narrow 10-storey building allows extensive cross-ventilation of air as well as penetration of natural daylight. And the rooftop is covered with photovoltaic solar panels, some 560 square meters in all, producing all the electricity needed for the building.

The Elithis Tower, home to the Elithis Engineering, is a first in France or anywhere, at least for an office building, but that is set to change, as France's tough energy standards kick in in just a few years. As of 2012 all new buildings must meet the low standard of 50 kWh per square meter per year, and by 2020, all new buildings in the country are to be positive energy. The implications for planning are considerable, of course. Adjusting codes and facilitating rooftop and neighborhood energy production remains a significant task. And there is much to be done to invest in the infrastructure and to put into place the incentives needed to move us toward this new global model. Rifkin endorses feed-in tariffs that exist in Europe and parts of Canada, and he envisions new housing finance instruments such as green mortgages that will encourage homebuyers to produce power.

Similar energy aspirations can be seen in Masdar City, the city planned on the edge of Abu Dhabi, in the UAE. Masdar aspires to be 'the world's first carbon neutral, zero waste metropolis' (Awad, 2009, p. 1) and will produce all of its energy from local renewable means. The headquarters building of the Abu Dhabi Future Energy Company, in Masdar City, seeks, like the Elithis Tower, to be a positive energy building. Much larger than Elithis, it incorporates a number of extremely interesting climate-sensitive design features. Special attention has been given to designing the structure to take advance of local climate. Most innovatively, the building integrates 11 wind towers, or wind cones, that utilize a natural stack effect to draw air throughout the building. The structure is also shaded by a large canopy and generous overhangs, with a rooftop also producing much power from photovoltaic panels.

Solar energy in particular will play an important role in the future city building and there are already many cities around the world that have been moving forward in support of solar. The Commonwealth government in Australia has provided funding for such city initiatives through a program called Solar Cities, and cities like Adelaide have developed strategies, and have already undertaken many projects (such as designating a portion of their down town as a solar precinct). Some cities such as Barcelona have adopted a solar ordinance that, in this case, mandate a minimum installation of solar thermal for hot water needs for new buildings or major renovations of new buildings. Portugal has followed suit and enacted similar requirements at the national level.

Cities like Freiburg have undertaken to subsidize and support solar installation and view solar energy as an important economic sector for the region. And there are now important examples of solar energy being integrated into Chinese cities, a positive development, as China has become a global leader in the production of photovoltaics. The city of Dezhou, for instance, located in the Shangong province in northern China, has become notable for its unusual commitment to solar energy and has become an inspiring example for other Chinese cities. There are an estimated half-million meters of installed solar hot water heating panels, for instance, and the city also boasts the world's largest solar building, a distinctive structure designed in the shape of a sun dial. While the energy demands and future projections for China are daunting, and currently some 70 percent of this energy is generated by burning coal, examples like Dezhou are encouraging and point the way to future urban priorities and practice. The Dezhou example also shows compellingly the job-generation potential of commitments to renewable energy (with some estimates that as many as 1.5 million jobs for Dezhou residents will be provided in this sector by 2020; Ra, 2009).

Resilient cities

Contemporary cities face a variety of serious environmental and economic challenges and shocks, and increasingly city planners and urban leaders argue for the need for cities to be more resilient. A term which emerged first in the ecological sciences and in disaster management, there is now considerable work in applying the concept to cities. The potential impacts of global climate change especially suggest that cities will need to develop the capacity to respond to and adapt to new and serious sets of conditions.

Much of the world's urban population sits along coastlines, for instance, where future sea level rise will be a significant problem. A recent study by researchers at the International Institute for the Environment and Development (IIED) estimates that about 10 per cent of the world's population, much of it urban, now resides in highly vulnerable low elevation coastal zones (LECZ), or locations less than 10 meters in elevation (a mere 2 percent of the world's land area) (McGranahan, Balk and Anderson, 2007). A specific look at the world's largest port cities finds many of them highly vulnerable. Nicholls et al. (2007) conclude that 40 million urban residents are already subject to a 100-year storm surge and that by the 2070s this could triple in number to around 150 million (and put assets of about $35 billion at risk). As urbanization and climate change progress together, in combination with subsidence, the exposure in many cities will grow dramatically (Mumbai's exposed population will grow from 2.8 million today to an estimated 11.4 million in 2070; Dhaka from 844,000 today to a projected 11.1 million). For cities in the developed world, property damage and exposure are greater concerns: Miami tops the list of 20 most asset-exposed cities, but New York and Virginia Beach are also there (as are a number of Chinese and Indian cities).

Cities like Miami and New York, in the United States, or London in the UK, will see significant expansion of the areas of coastal inundation in the future. How to plan for and adapt to these sea level rise predictions will be one of the most serious questions to be faced: will cities be able to (afford to) build new coastal defenses, or alternatively should cities engage in long-term shoreline 'retreat', steering urban growth away from flood-prone locations and relocating homes, business and infrastructure when opportunities arise? Should cities look for new and creative ways to design buildings and infrastructure that are better able to withstand future flooding, for instance following the Dutch who are experimenting with floating forms of homes in some areas.

Climate changes will create serious other problems for cities around the world, in addition to sea level rise. East-coast US cities, for instance, can expect dramatic increases in high temperatures during the summer months, further exacerbating air quality problems and creating a variety of public health problems.

Many cities globally face other serious natural disaster threats, notably earthquakes. The 2010 earthquake in Haiti has done much to raise global awareness, but few steps have been taken in the world's most vulnerable cities to seriously confront these threats. Roger Bilham, of the University of Colorado, has done extensive analysis of global vulnerability and concludes that the immense earthquake threat is largely being ignored, predicting that seismic events that result in a million deaths at a time are increasingly likely (Bilham, 2009). Particularly vulnerable are cities along the 'Alpine/Himalayan collision belt' (From Tokyo to London, with large vulnerable cities such as Dhaka, in between), and cities in close proximity to areas of high (tectonic) plate boundary strain. Bilham estimates that the total urban population living in these vulnerable cities is 1.2 billion. Bilham's work is a plea to incorporate seismic design into all future growth and development in these vulnerable cities: 'the current generation of earthquake engineers, city planners and political leaders have an unprecedented opportunity to render many of the world's cities resilient to earthquakes. It is clearly happening in the developed nations, but will it occur as successfully in the places under greatest threat?' (Bilham, 2009). Stronger seismic codes, and the political will to enforce them will be a key.

Glocal cities

Our understanding of the ecological and material flows of large cities, and their large ecological footprints, has increased greatly in recent years. Significant recent studies of the materials flow required to sustain cities such as London and Hong Kong, demonstrate, as mentioned above, the reality that modern cities exert tremendous resource pressures on the planet. Along with these pressures fall significant responsibilities to understand these impacts and take steps to minimize them. Localizing and regionalizing food, energy and other material inputs is one important strategy, but modern cities will by necessity still need (and want) to be connected to other cities and regions. *Glocalism* suggests the importance of the local, but also the need for responsible global sourcing and consumption and acknowledges the important positive role that cities might play on the global stage.

Parag Khanna, writing in the journal *Foreign Affairs*, declares the era of the nation over and makes a strong case that the city has emerged as the most important global unit: 'The 21st century will not be dominated by America or China, Brazil or India, but by the city. In an age that appears increasingly unmanageable, cities rather than states are becoming the islands of governance on which the future world order will be built. This new world is not – and will not be – one global village, so much as a network of different ones' (Khanna, 2010).

Glocal cities will understand that there are many new forms of global urban citizenship. There are new duties to forge material and resource flows that restore global environments rather than exhaust or diminish them, that the large urban populations with their significant buying power and demographic clout, can shape global resource management in many positive ways. As mentioned earlier, the city of New York is attempting to shift away from tropical hardwoods. Similarly, understanding a city's large inputs of food, say tropical fruits, suggests an obligation to foster fair trade and ecologically sustainable producers and organizations. Could large cities of North America support wood sourcing not only from Forest Stewardship Council (FSC) certified forests, but also from such places as the extensive network of community forests in Mexico (and elsewhere), supporting community, alleviating poverty and sending financial support for more sustainable, long-range landscape conserving modes of commerce.

One interesting development in the fair trade movement is the engagement of communities and towns. A number of cities (in the United States) are now participating in the Fair Trade Towns program, a kind of certification of these places, if they meet five key goals (including adoption of a resolution, and creation of a steering committee). The Fair Trade Towns campaign, based in the UK, reports almost 900 towns around the world participating. New Mexico aspires to be a fair trade state, and there may be a need for *fair trade cities*, which foster direct – more equitable ad sustainable – long-term relationships between consumers and retailers in large cities, and producers in places perhaps thousands of miles away.

Perhaps *glocal* cities must (will) establish new governance structures that recognize these new responsibilities to foster sustainability practices on the global scene. Do large cities need the equivalent position of secretary of state, perhaps building on and morphing the international trade and commerce offices that often already exist into something more about encouraging and supporting green and ecological practices around the world? Cities, especially larger cities, can establish programs and initiatives to share technology, knowledge and resources to help other global cities solve their problems. Large global cities may in the future, as a normal course, enter into treaties or long-term arrangements to help each other in confronting difficult environmental and sustainability issues. New Orleans, Rotterdam and Dhaka have much in common as delta cities, for instance, and Perth, Los Angeles and Cape Town share many of the challenges of water and managing (and protecting) an extremely biodiverse flora.

The urban way forward

That we have entered a global urban epoch is undeniable. How this urban transformation will turn out remains to be seen, and there are many potential outcomes. The best would be to harness the many positives of urban living – economic innovation, improved quality of life, advancements in democracy and development of the sector – while at the same time profoundly reducing the ecological footprint of urbanites. The potential to achieve many of the goals and aspirations of advocates of sustainability – carbon neutrality, a shift to renewable energy sources, reducing dependence on private automobility – have the greatest chance of success in an urban setting. As the above stories and descriptions of global practice demonstrate, there is already much happening. Virtually all of the new ideas we have for cities, and how to re-imagine cities of the future, have been at least partially put into practice. It is less about inventing new urban technologies or products or programs, than putting into practice those that have already been tested. It is about finding the political will, the popular (and political) support and the economic resources to put these many good ideas into global practice.

Notes

1. On the ways in which landscape architects are dealing with highly contaminated industrial sites, see Jorg Sieweke's article in this volume.
2. Alayne Reitman, EverGreen Cooperatives, Cleveland, from a film found at http://evergreencooperatives.com/, accessed July 2, 2013.

References

Awad, K. (2009) 'World's first carbon neutral city', found at: ISO Focus, found at: http://www.acec.ca/source/2009/May/CarbonNeutralCity.pdf, accessed July 2, 2013.

Beatley, T. (2010a) 'Biophilic Oslo', paper prepared for the Green Oslo research forum, Oslo, Norway.

Beatley, T. (2010b) *Biophilic Cities: Integrating Nature into Urban Design and Planning* (Washington, DC: Island Press).

Bilham, R. (2009) 'The seismic future of cities', *Bull Earthquake Eng*, September 2.

Copenhagen Post (2009) 'DONG Satisfied with Electric Car Tax Relief', Copenhagen Post, May 22.

Daseking, W. (2011) Personal Communication, May 2011.

Danish Architectural Centre (undated). 'Stuttgart: Cool City', found at: http://www.dac.dk/en/dac-cities/sustainable-cities-2/all-cases/green-city/stuttgart-cool-city/?bbredirect=true, accessed July 2, 2013.

Diaz, J. (2010) 'Honeycomb skyscraper has no internal structure, attracts giant killer wasps', found at: http://gizmodo.com/, accessed July 2, 2013.

Elgendy, K. (2010) 'Masdar headquarters: The first positive energy building in the Middle East', found at: http://www.carboun.com/sustainable-development/sustainable-design/masdar-headquarters-the-first-positive-energy-building-in-the-middle-east/, accessed March 7, 2011.

Embarq (2011) 'Mexico City launches third line of Metrobus BRT', found at: www.embarq.org/en/news/, accessed July 2, 2013.

Gallagher, J. (2010) *Reimagining Detroit: Opportunities for Redefining an American City* (Detroit: Wayne State University Press).

Gehl, J. (2010) *Cities for People* (Washington, DC: Island Press).

Jacobs, J. (1970) *The Economy of Cities* (New York: Vintage Books).

Khanna, P. (September 2010) 'Beyond city limits', *Foreign Affairs,* found at: http://www.foreignpolicy.com/articles/2010/08/16/beyond_city_limits, accessed July 2, 2013.

Lenoir, A., et al. (2010) 'Net zero building in France: From design studies to energy monitoring: A state of the art review'. IEA publication, found at: www.iea-shc. org/publications/downloads/Task40c-Net_Zero_Energy_Buildings_in_France_From_Design_Studies_to_Energy_Monitoring. pdf, accessed July 2, 2013.

McDonough, W. and M. Braungart (2002) *Cradle to Cradle: Remaking the Way We Make Things* (New York: North Point Press).

McGranahan, G., D. Balk and B. Anderson (2007) 'The rising tide: Assessing the risks of climate change and human settlements in low elevation coastal zones', *Environment & Urbanization*, 19(1), 17–37.

Nicholls, R. J. et al. (2007) 'The Ranking of the World's Cities Most Exposed to Coastal Flooding Today and in the Future', OECD Working paper number 1.

Owen, D. (2010) *Green Metropolis: What the City Can Teach the Country About True Sustainability* (New York: Riverhead Books).

Power, M. (2010) 'Bogotá's Ciclovia could teach Boris Johnson how to run a car-free capital', *The Guardian*, June 11.

Ra, A. (2009) 'China unveils world's largest solar office building', found at: www.solar-power-information-site.com/, accessed July 2, 2013.

Rifkin, J. (2009) *The Empathic Civilization* (New York: Jeremy P. Tarcher/Penguin).

Rifkin, J. (November 2010) Personal interview, University of Virginia.

SMIT (Sustainably Minded Interactive Technology, 2010) (undated). 'Solar Ivy is a solar energy delivery device that draws inspiration from ivy growing on a building', found at: http://solarivy.com/the_idea, accessed July 2, 2013.

Sustrans (undated) 'What is DIY Streets', found at: www.sustrans.org.uk.

Transport for London (undated) 'What are Barclay's Cycle Superhighways', found at: http://www.tfl.gov.uk/roadusers/cycling/15831.aspx, accessed July 2, 2013.

Travel and Transport Research Ltd, Urban Transport Benchmarking Initiative, Year Two, July, 2005, found at: http://www.transportbenchmarks.eu/pdf/Reports-y2/UTB2-A1.1-COMMON-INDICATOR-DATA-ANNEX.pdf, accessed July 2, 2013.

UN Habitat (2007) *State of the World's Cities 2010/2011: Bridging the Urban Divide* (Nairobi, Kenya: UN Habitat).

Part II
Germany's Energy Revolution

4
The German Green Party: From a Broad Social Movement to a *Volkspartei*

Arne Jungjohann

The history of the German Green Party is one of profound and frequent change. It is also a story of remarkable achievement. Compared to its sister parties in other European countries and beyond, the German Green Party has been the most successful in terms of both elections and government participation. The Greens have been able to pass important pieces of legislation, thus shaping and driving Germany's relatively progressive environmental, energy and climate policies. Looking back at their history, it is clear just how far the Greens have come. Despite facing profound challenges such as the loss of their political mandate in the first post-unification federal election in 1990, the Greens evolved from a heterogeneous political group into a goal-oriented, professional party capable of forming governments with different political partners at both the regional and federal level.[1]

This chapter provides an overview of the German Green Party's history and impact on environmental policy. The first section looks into the past and describes how, over the last 30 years, the Greens have come a long way in their development from a broad social movement comprised of various radical groups to an established, successful party.[2] The second section highlights how the Greens have shaped the environmental and climate agenda in Germany over the last decade by implementing the Renewable Energy Act, the phase-out of nuclear power plants and Environmental Tax Reform. Finally, the third section describes the current state of the German Green Party and provides readers with an outlook on the challenges ahead.

From a broad social movement to an established political party

Late 1970s: Social movements form the 'anti-party' party

At the time of their founding in 1980, the German Greens had neither a common identity nor a coherent ideology. Rather, the party brought together from across the country various social and political movements which considered themselves alternatives to the established political parties. This approach was expressed in the Greens' provocative self-description as the 'anti-party party'. Rejecting a political philosophy occupied solely with economic, financial and security issues, this new Green philosophy advanced post-materialist values and interests such as emancipation, peace, environmental protection, and women's rights and self-determination. Since none of the established political parties engaged these topics, the Greens became the political hub and motor for a wide range of social movements. Their constituency included the radical left, communists and communist splinter groups, emancipation groups for social minorities, Third World supporters, the emerging ecological movement and finally, the peace movement working for the disarmament of both sides during the Cold War. Although many different groups made up the movement, it was the ecologists, working on initiatives to meet local environmental concerns and stop large industrial projects such as airports and nuclear power plants, who had the idea to take the protests from the streets into the parliaments. After the initial success of these small political groups in local and state elections in the late 1970s, the desire to participate and collaborate in national collections sparked the founding of the official German Green Party on 12–13 January 1980 in Karlsruhe.

In accordance with its grassroots democratic ideals and origins, the German Green Party opted for a new type of party organization that allowed ordinary members to exercise permanent control over party officials, members and institutions. All internal party decision-making had to take into account that officials could be dismissed at any time. There were regulations for terms of office for senior positions, the separation of party office and parliamentary mandate, the principle of collective leadership and public committee meetings. Members of the national parliament had to step down after two years, even though the regular term lasts four years.

1980–1990: Entering the parliaments, fighting internal battles

In the 1983 national elections, the Greens entered the national political stage with 5.6 percent of the vote. Having thus cleared the 'five-percent threshold', a political mechanism that seeks to prevent fragmentation of the German party system, the party obtained 27 seats in the *Bundestag*, the lower house of the German parliament. A markedly unorthodox presence in appearance and aims, the party's journey from parliamentary outsider to junior partner in the 1998 government took place during the 16-year rule of the Christian Democratic Union (CDU) under Chancellor Helmut Kohl. These years of opposition strongly influenced the party's development, bringing it closer to the left-of-center Social Democratic Party (SPD), not just in policy but also in structure.[3]

As the Greens entered the various parliaments, internal divisions over different issues sparked intense debates and controversies. Increasingly separating into two different camps, party members disagreed about the fundamental nature of their role in the parliaments: should the Greens use the parliamentary stage to give voice to the extra-parliamentary movements that constituted their base? Or should they transform the oppositional 'anti-party' movement into a viable political player that functioned effectively within the established parliamentary system?

Whereas the party's fundamentalists (*fundis*) rejected any accommodation of the traditional party system, the more moderate proponents of realpolitik (*realos*) advanced the mid-term goal of entering governing coalitions at both the regional and national levels, with the aim of realizing the party's goals through legislation.[4] In 1985 the Greens formed their first formal governing coalition with the Social Democrats in the state of Hesse. Five years after their founding, the Greens had arrived as a serious player in Germany's political system. The coalition emerged against the wishes of the national party, however, which was dominated by fundamentalist ideas and objectives. The ongoing conflict between the *realo* strategy of participating in government and the *fundi* strategy of total opposition meant that there was no clear agreement as to what the Greens should do in parliament. Nuclear energy policy is an example of this divergence. After the Chernobyl nuclear disaster in 1986, the Greens demanded the instantaneous shutdown of all nuclear reactors. Since such a radical approach did not have a parliamentary majority at the time, the Hesse coalition between the Greens and the Social Democrats collapsed one year later.

1990–1998: The challenges of German unification for Germany and the Greens

The fall of the Berlin Wall in 1989 and the disintegration of the German Democratic Republic in the following year opened up a new set of challenges, for East and West Germany in general and for the Green Party in particular. How should the Greens, a party that strongly associated German nationalism with the Nazi past, deal with the possibility of German reunification? Moreover, what was the relationship of the leftist Green party to the East German Party of Democratic Socialism (the successor to the communist Socialist Unity Party) or the various East German civil rights groups?

Most Green Party members were skeptical of reunification due to concerns about the resurgence of German nationalism. As a result, the party entered the 1990 Bundestag campaign without a proper concept for reunification, except for the demand for a referendum on the new constitution. A campaign poster explicitly stated this lack of concern for reunification, noting that, 'Everyone is talking about Germany; we're talking about the weather!' Having misjudged the prevailing patriotic mood (or its power to focus public attention on other issues such as global warming), the West German party chapter won only 4.8 percent of the vote, thus failing to clear the five-percent hurdle. This failure to win representation in the national parliament posed a major handicap to the further development of the West German Green Party faction. That the Greens did not entirely vanish from the national scene was due to a partnership with the *Alliance 90,* an East German civil liberties group.

The Greens' unexpected failure in West Germany was an inglorious end to the first decade of Green Party politics. During this period, they had managed to establish themselves in the whole of West Germany, obtaining good results in both state and national elections. Policy development, however, had been hindered by exhausting internal debates over strategy. Even after 10 years of the Greens, the vast majority of voters still had no clear understanding of the party's position on economic and social issues or why they should vote for them other than on environmental issues. The crisis in 1990 and the newly reunited country that had threatened the very existence of the Greens offered the party an opportunity to start anew and clarify their positions. Now that the collapse of the German Democratic Republic had weakened the position of the utopian left and radical elements had departed, the time was ripe to develop a common basis for future political action.

In its most fundamental reform in 1991, the party unmistakably defined itself as an 'ecological reform party', finally putting to rest the question of whether the Greens were 'in' or 'out' of the political system. The arrival of environmentalists and civil rights activists from the East German Green Party confirmed this trend which was to mark the Greens' second decade. The intra-party power dynamics had shifted in favor of the *realos* (realists), in part because the experiences of the

East German movements had made them highly skeptical of socialist ideas. In addition, many East German party members were active in the Protestant churches. They tended to be Christian and conservative, thus favoring a more consensual and pragmatic political style. Other reforms included ending the two-year rotation of members of the parliament and replacing the three spokespeople with two. A political director was entrusted with party administration.

In order to have a chance of success in a reunified Germany, the Greens had to get a party structure up and running in the eastern part of the country quickly. Having fully merged with the small East German Green Party in 1990, they needed to think about working with the former East German civil rights movement *Alliance 90*, the parliamentary colleagues of the East German Greens in the Bundestag. It was for practical reasons and to their mutual electoral advantage rather than any real sense of common purpose and ideology that the two parties came together to form a new party, *Alliance 90/The Greens*, in 1993. Despite the party's double name, however, the much smaller Alliance 90 was unable in the following years to exercise much influence in the party, either in the appointment of officers or in policy formulation.

In the following years the Greens developed a goal-orientated program that allowed them to participate in state governing coalitions until they succeeded in returning to the Bundestag in 1994 with 7.3 percent of the vote and 49 parliamentary seats. If we view the period from 1990 to 1994 as the period in which internal party organization was revamped, it was the legislative period from 1994 to 1998 that saw a retreat from radical ideas and the development of a more strategic program. The parliamentary party, freed from the two-year rotation system and now more professionally organized, was properly able to steer green policy. The result was that in 1998 the Greens became part of a national government ruling coalition with the Social Democrats.

1998–2005: Governing the country: The 'Red-Green' coalition

The parliamentary Greens worked without any great controversy after 1994.[5] The party stood in the full gaze of public scrutiny as they had a realistic chance, in a coalition with the Social Democrats, to break the 16-year hold on power then held by the 'black-yellow' coalition between Christian Democrats and Liberals.[6] Despite a relatively modest showing at the polls with only 6.7 percent of the vote, Alliance 90/The Greens became part of a 'red-green' majority at the national level, joining the government with three ministerial portfolios: foreign affairs (and vice chancellor), environment and health. After only 18 years of existence the Greens had their hands on the levers of political power in one of the world's largest economies.

The faces of this coalition government were two men who would represent and shape its politics and style over subsequent years: Gerhard Schröder for the Social Democrats and Joseph 'Joschka' Fischer for the Green Party. Gerhard Schröder had been elected premier of the northern state Lower Saxony in 1990, and was re-elected in both the 1994 and 1998 state elections. In contrast to other Social Democrats, Schröder established a business-friendly reputation which helped him win 'new center' voters who previously mistrusted the political left. This brought him on collision course with the party's chairman, Oskar Lafontaine, who in protest against the new course resigned from all party posts in 1999. Schröder was party leader from 1999 to 2004. He represented a new generation of politicians who had little personal memory of World War II and felt at ease moving the seat of power back to Berlin, despite the city's association with Nazi Germany and the Prussian military state.

The leading figure for the Greens was foreign minister Joschka Fischer. He was one of the key protagonists in the development of the party. Fischer began his career with the Greens in the 1980s as a brilliant rebel most often seen in jeans and leather jacket. The public had a ringside seat as, on taking office as foreign minister and vice chancellor, Fischer transformed himself into a serious, responsible German representative on the world stage. His checkered past mirrors the history of the Federal Republic. His predilection for regularly changing his wives (to date Fischer has been married five times) and his ability to change his physical appearance by gaining and losing weight brought an air of glamor to the political scene that had been sorely missed in the long years of the previous government under conservative chancellor Helmut Kohl. In opinion surveys, Fischer's personal popularity rating was far above those of other politicians, which gave the Greens a significant advantage at election time. There was, however, a downside to Fischer's enormous popularity: if the Greens' leading personality decided to leave politics or should his popularity vanish, there was no other person to take his place. With Fischer in the party, it was impossible for others to project such a high profile or present themselves as possible successors. This 'Fischer Factor', combined with the reformed but still not entirely efficient organizational structure, was identified as a threat by political observers.[7]

During the four years of the red-green coalition from 1998 to 2002, the balance of power tipped in favor of the Social Democrats. This was not surprising given the fact that they won 40.9 percent of the vote and thus received a clear political mandate to form a government. Although the Greens' result (6.7 percent) was respectable for a small party, the Social Democratic Party was free to choose its coalition partner. This included, in theory, the conservative Christian Democratic Union and the business-friendly Free Democrats. Its strong bargaining position allowed the SPD to water down the Greens' more sweeping demands during negotiations for the coalition agreement and keep constant pressure on its junior partner. This included the two most important items on the Greens' domestic political agenda at the beginning of the legislative period: the reform of citizenship law and nuclear power. When during coalition negotiations the party was forced to cede its demand for an immediate shutdown of all German nuclear power plants, the stage was set for the first major conflict between the parliamentary group and its grassroots. For the majority of party members, the purpose of the Greens' participation in government was to put an immediate end to nuclear energy.

Another challenging question concerned the Greens' relationship to the peace movement. The decision by the North Atlantic Treaty Organization (NATO) to use air strikes to stop the expulsion of the Albanian minority from Kosovo confronted the coalition government with the question of war and peace. For the first time since the end of World War II in 1945, Germany had to decide whether to send its troops to fight in a foreign country. The conflict over this issue within the Green Party was an awakening for those who had thought that merely stating alternative ideas could change the world. The implications of making complex government decisions became shockingly clear when pacifist social movements, traditionally a reliable base of the party, suddenly began to protest against the Greens.

Considering the disillusionment that many party stalwarts experienced during the Greens' first term in national government, it was hardly surprising that the party had little electoral success during these years. Between 1998 and 2002, the Greens suffered heavy losses in all 15 state elections, causing red-green coalitions to disappear in several states. The Greens went into the 2002 Bundestag election, projecting the image of a responsible party tested in government. Against all odds, the red-green coalition pulled out a last-minute victory and squeaked back into office. It helped that heavy flooding along several major German rivers pushed climate change and environmental concerns to the front of the election campaign. The coalition also benefitted from Chancellor Schröder's announcement that Germany would not support the invasion of Iraq with troops. In the end, the Greens won their hitherto best national election result with 8.6 percent of the vote.

Although the SPD would continue to dominate the coalition in terms of electoral votes and parliamentary seats, many of the successful reforms passed under the red-green government had their origins in Green ideas and projects. Progressive social legislation included a new citizenship law enabling immigrants and their children to acquire German citizenship more easily and the recognition of homosexual partnerships. The coalition also successfully passed major environmental legislation, including the Ecological Tax Reform (1999), the Renewable Energy Act (2000) and the nuclear phase-out (2002). The latter was controversial among Green Party constituents, however, as the final version of the law stipulated a gradual phase-out through 2022, thus falling short of the Greens' goal of closing down the entire nuclear fleet at once.

Despite a strong showing in the 2002 national election, the Greens were unable to exert considerable influence on the second term of government. The major reason for this was Germany's weak economy and high unemployment rate. In response to this major economic crisis, Chancellor Schröder pushed for a major social reform package, called the 'Agenda 2010'. It included cuts to the welfare system (e.g. health care, unemployment benefits and pensions), lowering taxes and new regulation of the labor market. The changes introduced by Agenda 2010 marked the beginning of a painful time for the Social Democrats, as these changes were highly unpopular with their traditional voters. The Social Democrats took heavy losses in state and European Parliament elections between 2003 and 2005. The result was that the majority in the Bundesrat, Germany's upper house, moved ever more clearly in favor of the Christian Democrats. To pass the upper and the lower house of parliament, major bills had to be negotiated between Social Democrats and Christian Democrats. This arrangement weakened the position of the Greens, who had little influence on these negotiations.

At the same time, the Greens' potential party base was widening, due in no small part to the dramatic changes in party structure and policy implemented during their time in government. In the Bundestag elections of 2002, the former 'anti-party party' garnered support from parts of the electorate the Greens had previously been unable to reach. Their consumer and family policies, as well as the statesmanship of Foreign Minister Fischer, provided them with the profile of a dependable reform party capable of governing. As a consequence, the Greens began to pick up support from the conservative urban middle class, particularly in more affluent cities such as Hamburg, Berlin and Bremen.

The red-green coalition, however, was increasingly untenable both in the regional governments and on the national floor. The main reason was the unpopularity of Agenda 2010, Chancellor Schröder's reform of the German social welfare state and the labor market. When in May 2005 the SPD experienced an electoral disaster in North Rhine-Westphalia, Germany's most populous state, Schröder announced new national elections without consulting the Greens, thus effectively ending the red-green coalition prematurely. Even if the Greens had a good electoral showing, a renewal of the unpopular red-green coalition was out of the question. Since no other party signaled an interest in forming a governing coalition with the Greens, it was clear that the party would not remain in office. The challenge in the 2005 national election was to convince voters that the Greens would be an effective opposition party. When all votes were counted, they had won 8.1 percent. The Greens were now the smallest party in the Bundestag, but they had avoided the worst case scenario of not clearing the five-percent hurdle required to obtain seats in the federal parliament.

2005–2010: Recharging in opposition and adjusting to Germany's new party system

The 2005 Bundestag election brought significant change to the German political system. The poor showing of the two major parties (the Social Democrats at 34.2 percent and Christian Democratic Union at 35.2 percent) and the strength of the three smaller parties meant that none of the traditional coalitions (the Christian Democratic Union and the Free Democratic Party, or the Social Democrats and the Greens) was able to command a majority. The only possibilities were three-way coalitions between the SPD, Greens and FDP or between the Union, FDP and Greens; or a grand coalition between the Christian Democratic Union and Social Democrats. It was the latter option that eventually became reality.

The constellation of parties following the 2005 elections confronted the Greens with new challenges. They had used their years in office to revamp and modernize their policies and values. Although the Greens had taken steps to attract new voters, they needed to develop a strategy that would allow them to act effectively as an opposition party. In order to make the most of their new role in the five-party system, they needed to detach themselves from the SPD and explore mid-term opportunities for cooperation with the Christian Democratic Union and the liberal Free Democratic Party. With the departure of Joschka Fischer (who had pushed for a coalition with the SPD) soon after the election, Greens were free to reposition themselves.

A new policy campaign was required to bring together the parliamentary and ordinary party members. The Greens in the Bundestag favored a middle-ground strategy of moderate family, economic and immigration policies. The party conventions of 2006 and 2007, each a meeting of some 800 delegates of state and city chapters, wanted a much sharper political profile. They adopted environmental stances free of the compromises they had been forced to make during their time in government, along with more left-leaning social policies. Though previous policies helped the transition towards a low-carbon economy, the convention asserted that Germany was still far from providing a realistic response to the challenge of climate change. The Environmental Tax Reform and the Renewable Energy Act were seen as too modest given the challenge. Gearing their ecological and climate change ambitions towards a new 'radical realism',[8] the Greens proposed fundamental changes for a 'solar energy economy'.[9] Putting human dignity at the heart of their social policies, they left the red-green Agenda 2010 behind them. In all, the Greens established a profile, based on protecting the environment and self-determination, that would make them a suitable partner for both left and right coalition building. In the 2009 federal elections, the Greens reached their best-ever result with 10.7 percent of the vote, which translated into 68 Bundestag seats. This did not allow for a return to government, however. The conservative Christian Democratic Union, having won the election, decided to form a government together with their preferred coalition partner, the business-friendly Free Democrats.

Major successes of the German Green Party in environmental legislation

Like all member states of the European Union, Germany has to implement European environmental, energy and climate legislation as national law, such as a cap-and-trade system for greenhouse gases. Other policies, however, are developed at the national level. The Green Party has played a key role in Germany's push towards a low-carbon economy. Under the red-green coalitions from 1998 to 2005, Germany implemented several environmental policies. In particular, the Environmental Tax Reform (1999), the Renewable Energy Act (2000) and the Nuclear Phase-out (2002) are major milestones in the Green Party's first coalition government at the national level. As explained below, these projects were successful in terms of their environmental and economic impact as well as for the political profile of the party. In addition, in 2002 the Green Party managed to transfer the responsibility for renewable energy from the Ministry of Economics, a ministry with long-standing support for fossil energy, to the Ministry of Environment.

This move strengthened the representation of renewable energy interests within the government.

Nuclear phase-out

After a drawn-out debate among political parties and lengthy negotiations with nuclear power plant operators, in 2002 the Bundestag passed a nuclear phase-out law.[10] The law stipulated that the nuclear power plants in the country had to be shut down after an average life span of about 32 years. However, as part of the compromise, the utilities had a total nuclear electricity-generating budget and could transfer remaining kilowatt-hours from one reactor to another unit. Two reactors were shut down under the phase-out law (Stade in 2003 and Obrigheim in 2005). A third unit (Mülheim-Kärlich), which had been under long-term shutdown since 1988, was also closed for good. The construction of new nuclear plants was prohibited. In early 2011, before the catastrophe of Fukushima, 17 nuclear reactors had plant commissions, providing roughly 23 percent of the gross national power generation. This is less than in 1997, when nuclear power made up 30 percent of gross national power generation. Over the same time period, renewables more than made up for the drop in nuclear capacity, growing from 4 percent in 1997 to 17 percent in 2010.

Over the last few years, nuclear power repeatedly came under heavy public pressure. After a significant crisis in the German nuclear utility sector following a number of incidents at several power plants, the Ministry of Environment announced that electrical systems at nuclear power plants nationwide must be checked. Although the technical problems did not directly put the control of the reactors into jeopardy, the impression that, in spite of new management, the operators simply did not have control of the facilities grew. Opinion polls showed that almost three-quarters of those polled were in favor of the immediate closure of all older German nuclear power plants.[11]

In 2009, the center-right government under Chancellor Angela Merkel, pursuing a change in nuclear policy, extended the life of existing nuclear plants for up to 14 years beyond the life span attributed to them under the red-green phase out law of April 2002. The decision was controversial, reviving a debate that had calmed down in the previous years. In September 2010, approximately 120,000 people formed a 120-kilometer (74.5 miles) human chain between the nuclear power plants in Krümmel and Brunsbüttel in northern Germany to protest this policy. While the Christian Democrats held fast to their commitment to extend existing reactor commissions, they rejected the construction of new nuclear plants (for the time being) and agreed to the eventual phase-out of the German nuclear fleet. Chancellor Merkel justified her decision by designating nuclear energy a 'bridge technology' that kept emissions down until renewable energies had fully matured, thus easing Germany's transition into the solar age. Given the public distaste for nuclear power and considering the billions of dollars in additional profits utilities would make, the government required the nuclear industry to finance investments in renewables. For that purpose, the coalition implemented a new spent fuel tax on uranium.

On 11 March 2011, just two weeks before the highly important state elections in Baden-Württemberg and Rhineland-Palatinate, an earthquake and tsunami hit Japan, triggering a nuclear accident at the Fukushima Daiichi nuclear power plant. Faced with a forceful revival of the dormant anti-nuclear protest movement, Chancellor Merkel reacted swiftly. On March 15, Merkel abruptly put a three-month moratorium on her government's decision to extend the nuclear reactors' life spans. During this time, a commission was formed to review nuclear plant safety and the country's broader energy safety, and the country's seven oldest reactors were shut down, at first temporarily and then for good. In a dramatic reversal of her own nuclear energy policy, Chancellor Merkel then pushed aside resistance in her own party, especially from its industry wing, and supported the phase-out of Germany's nuclear reactors by 2022. Germany's political landscape has been profoundly marked by the Fukushima events.[12] At the time this chapter is being written, the political question remains not if, but how quickly Germany will phase out nuclear power.

The environmental tax reform

As any visitor to Germany will quickly notice that gasoline prices are significantly higher than in most other regions outside of Western Europe. In early 2011, a gallon of regular gasoline cost over $7 at the pump. This is more than double the average price of gasoline in the United States. The price difference is almost entirely due to higher excise taxes on mineral oil and other fuels that were intended to raise revenues. In 1998, the Social Democrats, together with the Greens, suggested introducing a set of tax reforms. The aim of the proposed reforms was to harness the multiple dividends invoked by advocates of green tax reforms, such as providing incentives for higher energy efficiency and reducing the tax burden on labor, thus contributing to a reduction in unemployment.

The Ecological Tax Reform Act passed by the red-green coalition in 1999 mandated increases in the tax rates on mineral oil and gas and introduced a new levy on electricity.[13] Taxes were increased annually over five years to give long-term incentives to switch to more efficient technologies and fuels. A number of exceptions, motivated by social and economic considerations, were initially included. These exceptions were intended to safeguard the competitiveness of the manufacturing, agricultural and forestry sectors, and to avoid undue hardship for lower income households. At the same time, revenues from the Ecological Tax Reform were returned almost fully to taxpayers in the form of a gradual reduction in social security contributions. In 2003, for instance, €16 billion raised through the tax reform were used to reduce payroll taxes, allowing pension contributions to be lowered by 1.7 percent. In lowering the labor costs, the environmental tax reform promoted employment and contributed to the creation of an estimated 250,000 new jobs. A smaller fraction of proceeds has been used to subsidize the deployment of renewable energy and the modernization of buildings. In the immediate years after the implementation of the reform, fossil fuel consumption declined between 3 and 4.5 percent annually.

The Ecological Tax Reform was a high-profile project pushed by the Green Party and the environmentally progressive arm of the Social Democrats. Germany's parliamentary system and its strict party discipline allowed the governing coalition to pass the reform against all resistance. Its implementation, however, was highly controversial. Like everywhere else, taxes are a politically sensitive issue in Germany. Opponents of the reform were quick to launch a media campaign against the proposed legislation. Given the complexities of its design, it was easy for critics to portray the tax reform as a mere increase in the fiscal burden and raise concerns over industry competiveness. Christian Democrats and Liberals wanted the reform repealed, leading observers to expect the project to become a casualty of partisan politics.

Despite significant opposition, the pressure to cooperate between the governing Christian Democrats and the Social Democrats from 2005 on, coupled with a yawning gap in the state budget, made a repeal of the tax reform unfeasible. Rather than repealing the reform, the grand coalition closed further loopholes in the energy tax. The growing budget deficit has made the current conservative government, previously an ardent adversary of environmental taxation, dependent on this tax revenue. Ironically then, it has been Merkel's center right coalition that introduced new taxes on uranium and aviation and slashed subsidies and tax exemptions for coal mining, commuters and energy use in industrial processes.[14]

Renewable Energy Act

The Renewable Energy Act (*Erneuerbare Energien Gesetz*, EEG) was enacted in April 2000. Embedded in a wider policy framework, it has been the cornerstone of Germany's success in renewable energy. Interestingly, its design was based on a feed-in electricity law of 1991, enacted by the government then led by the Christian Democrats to support farmers in Bavaria, who wished to sell electricity from their small hydro power plants into the grid. The original law did not spur new developments across the different renewable technologies. However, once its design was adapted and expanded, the impact was impressive. The law sets fixed rates for every kWh of renewable electricity exported to the grid for 20 years. It contains a priority grid access provision, which requires the grid operator to provide access to the renewable energy power plant. The rates differ according to renewable energy source, conversion technique, and plant size.[15]

Based on this law and several amendments, Germany has seen a remarkable expansion in renewables in the last decade.[16] The share of renewable electricity rose from 6 percent in 2000 to 17 percent in 2010, leading Germany to become the world's top country in terms of new capacity investments across all renewable energy technologies. By 2020 this share is expected to grow to 38 percent. The economic benefits of this development are impressive. New generation and conversion facilities have continuously risen over recent years, coming to a record high €20 billion in domestic investment and €21 billion in exports in 2009. By 2010, some 340,000 people worked the field of renewable energy, most of them in biomass, wind and solar. Renewable energy offset energy imports by at least €5 billion. Through the 'merit order effect' where renewable power offsets relatively costly power from old fossil fuel power plants, renewable electricity reduces overall energy costs in Germany by up to €4 billion annually.[17]

In contrast to the highly controversial projects of the nuclear phase-out and Ecological Tax Reform, the Renewable Energy Act has been supported by a broad parliamentary coalition. By 2010, all major German parties supported an industrial transformation toward a low-carbon economy. Constituent groups from progressive camps, such as the renewable energy industry, as well as conservative camps, such as farm communities, benefit from this approach. The understanding is that strong environmental policies drive ecological modernization and create new market opportunities. As an export-based country, Germany will be able to sell these innovations and solutions to a carbon-constrained world with high energy prices. The idea of greening the economy was first pushed by the Greens, but has been widely adopted by the other German parties.

The Greens on their way to a *Volkspartei* – opportunities and challenges

In 2010, protests erupted in Germany when Chancellor Merkel decided to extend the nuclear phase-out far beyond the previously set year of 2022. In Stuttgart, Baden-Württemberg's capital in Germany's south-west, tens of thousands of protesters marched against *Stuttgart 21*, a grand project to move the city's rail station underground. These demonstrators were not small in number or from the edge of the society, but spanned a broad spectrum in terms of ideology, education and age. Being in opposition on the federal level and in most states, the Greens sided with the protests and successfully articulated the frustration of many citizens with the political system, particularly the governing parties of the Christian Democrats and Liberals. The Greens greatly benefitted from this situation. In national polls at the end of 2010, 20 percent of Germans said they would vote for the Greens. As this chapter is being written, the Greens have climbed to a popularity level of 24 percent in national polls and are expected to win more seats and help appoint governors as part of ruling coalitions in several of the seven state elections of 2011. In Rhineland-Palatinate, the Greens tripled their votes in March 2011 from 4.6 to 15.4 percent and formed a coalition government with the Social Democrats as senior partner. In Baden-Württemberg, the Greens have made history by receiving not only 24 percent of the votes, but also appointing Winfried Kretschmann as the *Minister President* of Baden-Württemberg to head a green-red coalition with the Social Democrats. For the first time ever, the Greens are not the junior partner in a coalition, but actually head the government.

With the Greens winning state elections in this magnitude, commentators see a 'seismic shift' in Germany's political landscape.[18] To be sure, the accident of the nuclear power plant in Fukushima in March 2011 was a catalyst for the mobilization of green voters in Baden-Württemberg and Rhineland-Palatinate. However, considering the magnitude of the election results, the Green Party has reached a new dimension, even without the added effects of the Fukushima disaster. This is not only reflected by the results, but also the ability of the Green Party to attract voters from other political camps, such as conservatives and liberals. The party also performs much better in constituencies that it originally had trouble with, such as the unemployed and blue-collar workers. The substance of the Green Party's success is underscored by the rapid rise in membership over the last several years, from 45,000 in 2007 to more than 55,000 in

2011.[19] Double-digit results in elections will not be exceptional, but will become the norm (with the exception of parts of Eastern Germany). It is becoming clear that the Greens have entered yet another stage in their evolution. They have established themselves as the third party behind the Christian Democrats and Social Democrats, clearly in front of the Liberals and the Leftist Party. Political analysts see indications that the party is on the rise to a *Volkspartei*, a 'big tent' party.[20]

There are several reasons for the Greens' recent success and popularity. Unlike other parties, the Greens did not have to adjust their position after the nuclear accident of Fukushima. In addition, various other reasons explain this new stage in the development of the party:

1. Green core identity: Over the last few decades, green issues have moved to the political center in Germany. Values and political principles advanced by the Greens for more than 30 years, including combating climate change, phasing out nuclear power, striving for an ecological modernization of the German economy, embracing activist politics and promoting multiculturalism, feminism and gay rights are becoming mainstream. Organic food markets, renewable power, bicycling and car-sharing are lifestyle forms with growing support. Over the years, the Greens have remained true to many of their ideals, though policies have been adapted and their style has changed. They have created a green trademark, and voters know what they stand for. The Greens have sharpened their profile with a clear stand against nuclear power, genetically modified organisms, discrimination, social exclusion and infrastructure projects like coal plants. In addition, the party has developed positive visions and concrete concepts for a renewable energy economy, more transparency across society, the rights of women and children, fair access to education, social inclusion and better data protection. It has been a while since the Greens traded sandals or tennis shoes for suits and became the country's new establishment. The party has earned the public trust by staying true to their principles and beliefs.
2. Professionalization and leadership: Over the course of 30 years the party has changed and matured in terms of both membership and voters. The Greens have lost their radical roots. Leaving behind destructive fights between the *realos* and *fundis,* the Greens have demonstrated predictability, unity, and discipline in recent years. The party's operational structure has been professionalized over the years. This approach has paid off. The issue gap between political camps within the party has become much smaller, and the party's course

is mostly discussed behind closed doors, out of the public's view. In contrast to other parties, the Greens have cooperative leadership. When all other parties, with the exception of the Leftist Party, campaigned with one top candidate in the 2009 national elections, the Greens presented two. Now, two speakers of both the party (Claudia Roth and Cem Özdemir) and the parliamentary group (Renate Künast and Jürgen Trittin) represent the Greens in the public. In addition to the federal level, more Greens have taken on responsibility in the *Länder* (states) as Ministers in coalition governments. As head of the government of Baden-Württemberg, Winfried Kretschmann, the first 'Green Governor', takes a prominent role as a national figure and in coordinating the Greens in the *Bundesrat* (the upper chamber). Party members have directly elected two nominees for Germany's 2013 federal election. This was the first time ever after World War II a German party has asked its ordinary members to vote on the nominees. As expected, a top-ranking member, Jürgen Trittin, earned the nomination, but so did Katrin Göring-Eckardt in an upset. The parliamentarian and chair of the synod of the Evangelical Church defeated popular female challengers, like party head Claudia Roth and the other co-chair of the parliamentary group, Renate Künast. In comparison to other parties, the Green Party has a diverse, decentralized leadership. Depending on the level of cooperation, this can play out as advantageous to the party. Different Green Party leaders address different constituents and different themes with high personal credibility. Internally, their cooperation helps to enforce a joint decision-making process among the political wings within the party. Externally, teamwork allows the party to spread the burden of election campaigns on several shoulders, increasing visibility and showing presence across the country. This simultaneously makes them less vulnerable to poor performances of a single leader. The Greens have also won recognition for the depth of political talent on the sub-national level. Personalities such as Hessian party leader Tarek Al-Wazir, Schleswig-Holstein's *Energiewende* Minister Robert Habeck and Mayor of Tübingen, Boris Palmer, have honed political skills and voter appeal that often put them in the position of leader of the opposition (or in the case of Tübingen even in charge of the town hall) despite not being the largest opposition party.

3. Greening the economy: In comparison to its earlier days, the Green Party has widened its issue portfolio. The major drive to form the Greens in the late 1970s was a growing sensitivity to a post-materialistic approach to environmental responsibility, women's emancipation and human

rights. The issues of environmental protection, fiscal responsibility and greening the economy resonate with voters from the political center. The party has shifted to the political middle and also attracts voters disappointed with center-right parties, even including some conservative voters. Some argue that Green is the new black.[21] Overall, the Greens' middle-aged, university-educated constituency is growing as a proportion of the electorate. Over the last few years, the Greens have added a distinct economic component to their portfolio, pushing the idea of driving the economy with ecological innovations (see Figure. 4.1). The success of renewables as a major engine for economic and job growth in Germany has proven them right and rewarded them with more votes. Farmers, small businesses, craftsmen, engineers and architects, among a variety of other occupations, benefit from a greener economy. While old economic models have lost credibility, the green economy is robust and highly competitive. Based on this experience, the Greens are transferring this approach of a major industrial transformation to other sectors of the economy like transportation, buildings and chemistry.

Figure 4.1 Green election posters, 1980 and 2009

Note: *Left*: A Green Party poster for the 1980 national election: 'Ecological awareness, social consciousness, grassroots democracy, non-violence'. *Right*: Poster for the 2009 national election: 'Jobs, jobs, jobs. The path out of the [economic] crisis is green'.

Source: *Left*: http://bildungsserver.berlin-brandenburg.de/fileadmin/havemann/index.php/glossar/D.html. *Right*: http://www.gruene.de/einzelansicht/artikel/unsere-wahlplakate.html.

Outlook

The Greens know from their past that good polls may not automatically turn into votes. The nuclear accident of Fukushima in March 2011, however, catapulted them into a position of high credibility, leading to a position as the head of a state government for the first time. How long this trend will last remains to be seen. However, there are strong indications that this is not a temporary success, but rather an indication of substantial, long-term growth of the party.

Over the decades, the Greens have managed to become a reliable coalition partner in different combinations. The alienation between Greens and Conservatives on the nuclear phase-out seems to be dying down as the CDU was reversing its nuclear policy in the aftermath of the Fukushima accident. The ability to form a coalition with both sides of the political spectrum demands flexibility within the party, however, and does not come without disappointment to some of their voters.

As the party is entering more and more state governments, the Greens' general role in the political debate will change yet again. Over the last few years as an opposition party on the national level, the Green Party grew closer to the protest movements against infrastructure projects, such as Stuttgart 21, and the extension of nuclear power plant commissions. As long as the Greens were in opposition to the federal and state government, they could afford to march on the streets with the protesters. Returning to power, however, the Greens will have to implement projects in government that they might be able to alter a little, but not completely prevent. They will not be able to head both the government and the protests on the streets. In the beginning of 2013, the Greens are governing as coalition partners in 6 out of 16 *Länder*. This expansion of governing power brings also an expansion of resources as more professional staff switches from the party and parliamentary groups to a career path in government. At times, this can be a constraint to the party structure, but it will bring benefits in the future.

The new role of not only being a junior coalition partner, but leading a government such as the one in Baden-Württemberg, brings along more responsibility and challenges. As a small coalition partner, the Greens could focus on being the corrective element in the coalition. Leading a government, however, demands more responsibilities. Small parties can take clear stands in conflicts, because they don't have to worry if a majority of the population rejects their position. The bigger a party gets, the more it has to take into account the will of the majority. Big parties have to moderate conflicts for greater compromises, a stressful job that does not always come without contradictions. Looking ahead, such compromises could alienate some of the Greens' own constituents. The ecological transformation of industrial society requires dramatic changes and sometimes painful compromises. Additional gas and hydropower plants, high-voltage power lines and the exploration of underground storage for carbon dioxide are needed for the transition to a low-carbon economy. To master this challenge, the Greens will have to find a path forward that includes a moderation of all interests and sets expectations for their voters. The Greens still have to demonstrate that they can master these upcoming potential challenges without disappointing their voters.

If the current trend of five parties in the parliaments continues, varying coalitions will be the norm. So far, this has been an advantage for the Greens. While the Liberals have been politically and personally tied to the Christian Democrats, the Leftist Party sees its only natural ally in the Social Democrats, even if this is often an uneasy alliance. Coalitions between Social Democrats and Liberals have occurred only rarely; coalitions between the Christian Democrats and the Leftist Party only on the local level.

With a general openness and political ability to form coalitions with both Christian Democrats and Social Democrats, the Greens have more options and depend less on the performance of only one potential coalition partner. As a consequence, the Greens could find themselves more often in the position of the kingmaker. As of early 2013 the German Green Party has established itself clearly as the third force in Germany's political system. In the upcoming federal election, the Green Party hopes to join with the Social Democrats (SPD) to dissolve the government led by Chancellor Angela Merkel. It remains to be seen if the growth of the Greens continues.

Notes

1. The Federal Republic of Germany consists of 16 states. The parliament of a federal state or *Land* is called *Landtag*. The three city states call their parliaments *Abgeordnetenhaus* (Berlin) or *Bürgerschaft* (Bremen and Hamburg).
2. The first chapter is a condensed and edited version of the text Haas, 'The German Greens' (2008), http://www.boell.eu/downloads/GREEN_IDENTITY_UK_web.pdf, accessed 1 March 2011.
3. In Germany, political parties that receive less than five percent of the vote are removed from the count, and those votes are spread proportionally across the parties that received at least five percent. The five-percent rule was implemented based on experience in the Weimar Republic, when large political camps were splintered across a number of small parties, eventually allowing the Nazis to become the largest single party. For an overview on

German parties, see Klaus Dahmann (2009) 'Political parties form colorful spectrum in Germany', http://www.dw-world.de/dw/article/0,,4541120,00.html, accessed 4 June 2011.

4. Because the German political system is a multi-party system based on the principle of proportional representation, it is uncommon that one party wins the absolute majority. Forming a minority government is possible; however, the more conventional solution is forming a governing coalition with either the SPD or CDU, and one smaller coalition partner. Until the appearance of the Greens in 1980, the junior partner in such a government was the business-friendly Free Democrats. The year 2007 saw the founding of the Left Party, a merger of the successor party of the former East German Communist party and social democratic splinter groups.
5. See Thomas Poguntke (1999) 'Die Bündnisgrünen in der babylonischen Gefangenschaft der SPD?' in Oskar Niedermeyer, ed., *Die Parteien nach der Bundestagswahl 1998* (Opladen: Leske + Budrich).
6. In Germany, parties are assigned with individual colors: The Christian Democrats are black, the Social Democrats are red, the Greens are green, the Liberals are yellow and the Leftist are dark red. Coalitions are often named in terms of their colors, such as a black-yellow or a red-green coalition.
7. Joachim Raschke (2001) *Die Zukunft der Grünen. So kann man nicht regieren* (Campus-Verlag, Frankfurt/Main).
8. Bündnis 90/Die Grünen (2006) 'Für einen radikalen Realismus in der Ökologiepolitik'. Decision by the 26th *Bundesdeligiertenversammlung* (convention of delegates) in Cologne.
9. Bündnis 90/Die Grünen (2007) 'Klimaschutz ohne wenn und aber: auf dem Weg zur solaren Gesellschaft'. Decision by the 27th *Bundesdelegiertenversammlung* (convention of delegates) in Nuremburg.
10. Act on the Peaceful Utilization of Atomic Energy and the Protection against Its Hazards (Atomic Energy Act – AtG), http://www.bmu.de/english/service downloads/doc/3231.php, accessed 12 June 2012.
11. Mycle Schneider et al. (2009) 'The World Nuclear Industry Status Report 2009', http://www.bmu.de/files/english/pdf/application/pdf/welt_statusbericht_atomindustrie_0908_en_bf.pdf, accessed 12 June 2012.
12. Mycle Schneider (2011) 'The World Nuclear Industry Status Report 2010–2011. Nuclear Power in a Post-Fukushima World', http://www.worldwatch.org/nuclear-power-after-fukushima, accessed 12 June 2012.
13. Michael Mehling (2000) 'The ecological tax reform in Germany', *Tax Notes International* 26, 871–878, and in this volume.
14. Green Budget Europe (2010) *GreenBudgetNews*, 27 December, http://www.foes.de/pdf/GreenBudgetNews27.pdf, accessed 12 June 2012.
15. J. Farrell (2009) 'Driving the Economy with Renewable Energy Policy that Works' (Heinrich Böll Foundation, North America), http://www.boell.de/climate-transatlantic/index-129.html, accessed 12 June 2012.
16. German Federal Ministry of the Environment, Nature Conservation, and Nuclear Safety (March 2010) 'Development of Renewable Energy Sources in Germany2009', www.erneuerbare-energien.de/files/pdfs/allgemein/application/pdf/ee_in_deutschland_graf_tab_2009_en.pdf, accessed 12 June 2012.
17. Renewable Energy Network 21 (September 2010) 'Renewables Global Status Report 2010' (Paris), http://www.ren21.net/Portals/97/documents/GSR/REN21_GSR_2010_full_revised%20Sept2010.pdf, accessed 12 June 2012.

18. Charles Hawley (2011) 'A seismic shift in Germany's political landscape', *Der Spiegel*, 28 March.
19. Steffi Lemke (2011) 'Mehr innerparteiliche Demokratie wagen', 14 April, http://steffi-lemke.de/allgemeines/2011-04-14/mehr-innerparteiliche-demokratie-wagen, accessed 12 June 2012.
20. Lothar Probst (2011) 'Die Grünen auf dem Weg zur Volkspartei?' http://www.boell.de/demokratie/parteiendemokratie-die-gruenen-auf-dem-weg-zur-volkspartei-11767.html, accessed 12 June 2012.
21. Paul Hockenos (2011) 'Green is the new black', *Foreign Policy*, 1 March.

5
Germany's Ecological Tax Reform: A Retrospective

Michael Mehling

After months of heated political debate, the German government adopted a new Energy Concept in September 2010, setting out a broad framework for federal energy policy until 2050. Elaborated by the ruling center-right coalition, this document aims at turning Germany into one of the 'most energy-efficient and greenest economies in the world while enjoying competitive energy prices and a high level of prosperity'.[1] In line with a campaign pledge set out in the government's coalition agreement, the Energy Concept defines ambitious objectives for the medium and longer term.[2] Energy pricing through taxes and charges has traditionally held a prominent position in the German energy policy mix, and will also be central to achieving the targets adopted with the new Energy Concept.

Mineral oil and other fuels had already been subject to a system of excise taxes dating back to prewar Germany. It was not until the late 1990s, however, that energy taxation also became a vehicle for Germany's green agenda. Before that, environmental benefits of taxation had been no more than an indirect, if not unwanted, side effect in the process of creating public revenues.[3] As part of its electoral campaign, however, the center-left coalition of Social Democrats and Green Party that came to power in 1998 expressly pledged the introduction of new fiscal instruments to reduce the tax burden on labor and shift part of it to energy consumption. In doing so, it sought to harness the multiple benefits touted by advocates of green taxes: greater flexibility and cost efficiency than traditional regulation in achieving environmental policy objectives, a robust and economically effective incentive to develop innovative clean technologies, and the ability to raise revenues for public investments or tax cuts in other areas, such as labor costs.[4]

Yet the proposed tax reform was by no means uncontroversial: from the outset, it encountered public opposition triggered by rising prices for crude oil and concerns over industrial competitiveness. Initial resistance was, in fact, so great that many observers already expected the entire effort to end up a casualty of partisan politics. And yet, nearly a decade and several governing coalitions later, the ecological tax reform remains in place. What were its main characteristics, how was it adopted and what have been its principal impacts? Drawing on an historical retrospective, this chapter attempts to formulate some answers to these questions.

Introducing a price on energy: Good policy and good politics?

The Ecological Tax Reform Act of 1999

As stated earlier, the plan for an ecological tax reform was put forward by the ruling coalition. For a policy proposal to become formally enacted as an act of German parliament, however, a legislative initiative needs to pass both houses of parliament in what is typically a rather complex process. On 3 March 1999, the German *Bundestag*, or lower house of parliament, launched the first stage of the ecological tax reform by adopting the corresponding bill[5] by a narrow vote of 332 to 299 representatives. After obtaining approval by the Federal Council, or *Bundesrat*,[6] the Ecological Tax Reform Act entered into force on 1 April 1999.[7] Consisting of an act with merely three articles, the first stage of the Ecological Tax Reform introduced a separate law for the taxation of electricity and an amendment of existing legislation on mineral oil taxes. Both steps were specifically aimed at encouraging changes in polluting behavior and, thus, at improving overall environmental quality.[8] The revenues from environmental taxation were allotted to reducing Germany's traditionally high, non-wage labor costs, which include the employers' share of pension contributions, unemployment insurance, health insurance and other social security contributions. With the first year's revenues, an estimated 11.3 billion deutschmarks (approximately 5.8 billion euros),[9] non-wage labor costs were reduced by 0.8 percent to 19.5 percent of wages.[10] With hiring rendered less expensive, the ecological tax reform was to help promote employment. The government also forecast a beneficial effect on the structure of German trade and industry, maintaining that greater usage of the innovative potential offered by energy conservation measures would reinforce the country's leadership position in that field.[11]

The two principal features of the Ecological Tax Reform Act were a levy on electricity of 2 pfennigs per kilowatt-hour and an increase in mineral oil taxation by 6 pfennigs per liter on fuel, 4 pfennigs per liter on heating oil and 0.32 pfennigs per kilowatt-hour on gas. Massive protests from interest groups and general concern for the competitiveness of German industry, however, prompted the legislature to include a number of exemptions and abatements. Accordingly, the manufacturing, agricultural and forestry sectors were to benefit from a reduced tax rate on electricity if their yearly tax burden rose above 1,000 deutschmarks.[12] Further, manufacturing industries were allowed to claim a refund in the event that their tax payments significantly exceeded savings incurred by labor-cost cuts. For the manufacturing sector, in particular, this ensured that the Ecological Tax Reform would not introduce net additional costs and hence compromise industry's ability to compete in international markets.

As an incentive to promote its goal of improved environmental quality, the Tax Reform Act also provided for a full exemption for electricity obtained from wind energy, solar energy, geothermal energy, biogas, biomass and hydroelectric plants. Further abatements, among them a generally reduced rate on electricity used for public transportation and railway traffic, were included with regard to the unique significance of the affected sectors.[13] Proposals aimed at a general tax exemption for energy-intensive industries, however, had to be abandoned in face of pressure from the European Commission.[14]

The electricity tax introduced with this reform has been generally levied whenever an end-user draws electricity supplied by a provider through the transmission grid; the second, and less frequent, charge arises when a provider draws electricity for its own use. Electricity sold to a re-seller may, thus, be exempt from the tax. Although the tax is payable by the provider, not the end-user, the resulting cost burden is typically passed on to ratepayers.[15] As a consequence, the latter bears the actual burden of the tax, justifying the tax abatements and refunds for the manufacturing, agricultural and forestry sectors as well as renewable energy and public transportation.

A number of abatements were also included with regard to mineral oil taxation. Designed to prevent an excessive tax burden on certain industries as well as on the forestry and agriculture sectors, a provision in the Tax Reform Act froze existing rates on heating fuel and gas used in manufacturing processes or for the production of electricity. Combined heat and power systems with a minimum operating efficiency of 70 percent, for instance, were completely exempt from mineral oil taxation, whereas comparable systems with an efficiency of 60 percent or more are exempt from all rate increases. And finally, tax benefits for gas-powered vehicles were affirmed for an interim period.[16]

The Ecological Tax Reform Continuation Act of 2000

Draft legislation for subsequent stages of the tax reform was submitted by the federal government on 25 August 1999.[17] With the ruling coalition short of an absolute majority in the Upper House, and confronted with setbacks in several regional elections, it remained uncertain until the very end whether the bill would be able to surmount all obstacles on the path to parliamentary approval. The hearings were, not surprisingly, accompanied by heated disagreement.[18] In the Upper House, the opposition parties launched several petitions with requests for an amendment of the bill. At an earlier date, moreover, the minority caucus of the Liberal Democratic Party (FDP) had submitted an alternative proposal for an ecological tax reform, hoping to stall any efforts of the federal government and the ruling coalition.[19] Notwithstanding strong support from members of both opposition parties, this counter-proposal failed to receive the required number of votes.[20] Two major petitions for an amendment of the bill were likewise rejected during the parliamentary proceedings of the Federal Council.[21]

Overcoming this fierce opposition, the Ecological Tax Reform Continuation Act[22] eventually passed both Houses of Parliament with a narrow majority of 331 to 285 votes,[23] and took effect as scheduled. Thus, on 1 January 2000, the second of five designated stages of the tax reform incurred rate hikes in mineral oil and electricity taxation, picking up where the first stage had left off. Consisting of four draft bills, this package included a sizeable amendment to the initial Ecological Tax Reform Act of 1999. Compared with the previous year, the rate hikes in 2000 were predicted to yield additional revenues of 5.1 billion marks (2.6 billion euros) in 2000, 10.5 billion marks in 2001, 15.8 billion marks in 2002 and 21.2 billion marks in 2003.[24] Altogether, the federal government expected to cut non-wage labor costs by a further percentage point.[25] Indeed, the second stage of the tax reform already allowed for a reduction of pension fund contributions by 0.2 percent, bringing down the applicable rate to 19.3 percent.[26] Larger cuts were ruled out in view of the demographic developments at the time, which added to significant structural deficiencies in the prevailing social insurance scheme.

Notwithstanding the ample proportions of the second stage, the structure of the original Tax-Reform Act – including the rules on abatements and refunds – was essentially retained. The government had taken heed of its opponents, and was signalizing its intention to safeguard the international competitiveness of German industry.[27] Accordingly, the central feature of the new act consisted of a series of consecutive and, as the government emphatically pointed out, moderate rate hikes.

From 1 January 2000, onward, the electricity tax was subjected to an annual increase of 0.5 pfennigs per kilowatt-hour, amounting to a total increase of 2 pfennigs by 1 January 2003. As an additional incentive for the usage of renewable energy sources, the tax exemption for electricity thereby obtained was no longer limited to power plants with a maximum output of 5 megawatts.[28]

The new act also provided for a rate hike on mineral oils, including diesel and gasoline. The rate was augmented by 6 pfennigs a year for a total increase of 24 pfennigs per liter over the following four years. Tax rates on natural and liquefied gas used for vehicle fuel were adjusted to the rates on mineral oil. The rates on heating fuels, however, were to remain unaltered for social reasons, as it was feared that low-income households would be hit hardest, and residential heating is not a form of energy usage that can be easily altered.[29] In order to reduce the burden on German authorities, the new rules also applied a uniform rate on heating fuels, therefore making no distinction between fuels used for heating purposes and fuels used to obtain electricity.

Environmental considerations prompted the inclusion of an additional hike in taxes on high-sulfur fuel, which took effect on 1 November 2001. Moreover, the decentralized production of electricity was given further impetus by extending the exemption formerly reserved for combined heat and power plants to all plants operating with joint gas and steam turbines at an efficiency ratio of 57,5 percent or more, provided those plants were put to use between 31 December 1999 and 3 March 2003. Public transportation continued to benefit from a substantial abatement on the normal rates.[30] And finally, a sizeable portion of the revenue from the ecological tax reform, 200 million marks, was used each year to support the usage of renewable energy sources.[31]

The Energy Tax Act of 2006

In 2005, the center-left coalition of Social Democrats and the Green Party was replaced by a center-right coalition between the Social Democrats and the Christian Democratic Party (CDU/CSU). This brought a slightly changed political dynamic to the discussion on energy taxation. On 29 June 2006, the lower house of parliament adopted an act to modify the taxation of energy products and amend the Electricity Tax Act. Approved by the Federal Council on 7 July 2006, this act entered into force on 1 August 2006[32] and deferred further rate hikes for energy products as set out in the coalition agreement between the SPD and the CDU/CSU under Chancellor Angela Merkel.[33] Passed to fulfill this political commitment, the act also transposed a directive of the European Union[34] and responded to a judgment of the European Court of Justice of 29 April 2004, which had censured Germany for applying the definition of 'heating fuel' too narrowly.[35] To this end, the act placed energy taxation in Germany on an entirely new legal basis. Aside from revising the rules on electricity taxation, it repealed earlier legislation on mineral oil taxes and introduced a tax on coal and lignite used for heating purposes. All that was incorporated in a uniform Energy Tax Act (*Energiesteuergesetz*), setting out a common fiscal framework for energy products through harmonized definitions, taxation rules and exemptions. Application of the new act is facilitated by an ordinance guiding the implementation of individual provisions.[36]

Altogether, the adoption of this comprehensive act clearly marked a new stage in German energy taxation. On closer view, however, the changes incurred by the new act are not as dramatic as they may have initially seemed. Consisting of 67 provisions, the Energy Tax Act retained large parts of the earlier Mineral Oil Tax Act it was set to replace. With rates largely left unchanged, the amendments have been largely structural in nature. For instance, in keeping with requirements under European Union law, all energy products used for electricity generation are now equally exempt from taxation. Gas is no longer taxed at the point of insertion into the distribution grid, but rather with its supply to end users, and preferential treatment of liquid gas and natural gas will continue until 2018. Simplified rules on abatements for combined heat and power generation and generally favorable rules on efficient gas and steam turbines, moreover, were seen to improve the prospects for greenhouse gas reductions in the power sector.

Genuine changes have been largely confined to the introduction of new taxes on coal, lignite, coke and biofuel. For the first time, coal, lignite and coke used for the generation of heat were subjected to an energy tax of 33 cents per gigajoule of energy content, although private households were again exempt from payment until 2010 out of concern that rising heating bills would affect low-income households hardest, with few opportunities for behavioral change to avoid paying the tax. Likewise, following lengthy negotiations, a compromise was reached to implement a system of gradually increasing rates for previously exempt biofuels, a measure that would complement a mandatory quota scheme for conventional fuel. Until the end of 2007, vegetable oil remained exempt from taxation, while pure biodiesel was subjected to a tax of 9 cents per liter. Between 1 January 2008 and 31 December 2012, taxes on vegetable oil and biodiesel are being successively raised until they reach a level of 45 cents per liter in 2012. Biodiesel used as an additive to conventional fuels for attainment of the mandatory quota has been taxed at 15 cents per liter from 1 January 2007 onwards. Biofuels used in agriculture and forestry, however, have remained exempt from taxation, as have all fuels – including biofuel – used in public transportation. Tax waivers have also been retained for a number of energy intensive sectors, such as the production of cement, glass and metals.

Assessing the German ecological tax reform

Although comprehensive, the ecological tax reform described in the preceding section should not be misconstrued as an indicator of unconditional acceptance in Germany. Throughout the legislative process, numerous stakeholders voiced serious concerns about the reform project. Opponents voiced their fear of a negative impact on the German economy and argued that the new fiscal burden would increase the confusion and uncertainty already prevalent in German tax law while failing to achieve the original goal of reducing environmental pollution.[37] Such resistance was summarized in a recommendation submitted to the Federal Council by the Parliamentary Committee on Economic Affairs.[38] In this document, the committee explicitly accused the proposed tax reform of placing an unnecessary strain on the German economy, while achieving no progress in the field of environmental protection. Specifically, it forecast a further decline in the global competitiveness of German industry and denounced an inequitable distribution of the ensuing tax burden, which, as the committee held, would place certain members of society at a substantial disadvantage. Rather than produce the expected 'double dividend', the ecological tax reform would cause certain industries to leave the country. At the same time, the cost of energy was feared to reach untenable levels in the eastern part of Germany, an area still struggling with the economic burden of its past. In spite of a complex abatement scheme, the committee also forecast an increase in expenses for the public transportation sector, an effect which, it added, would prove counterproductive to the goal of environmental protection. Further arguments relating to the structure of the bill and its prospects of inviting similar measures from neighboring countries were, finally, brought forward to solidify the opposition against the tax reform.

How do these dire projections hold up against actual experience a decade later? In retrospect, Germany has fared quite well despite the increased energy costs incurred by the ecological tax reform. As intended, between 1999 and 2003, the tax reform resulted in a gradual increase of energy costs. Over this period, the rates for gasoline and diesel grew by

3 cents per liter each year, the rates for heating oil by 2 cents per liter, and the electricity tax, which had first been introduced on 1 April 1999 with 1 cent per kilowatt hour, by 0.26 cent per kilowatt hour. Liquefied natural gas, coal, lignite and coke used for transport or heating purposes as well as previously exempt biofuels have also been subject to the energy tax. Overall, however, the fiscal burden resulting from the environmental tax reform has been moderate compared to already existing taxes: for instance, only 15 cents of the 66 cents currently charged in taxes on every liter of gasoline are a result of the ecological tax reform, with the far greater share originating in the excise taxes already imposed prior to 1999. Altogether, the share of ecological taxes in the tax burden only rose from 5.2 percent in 1998 to 6.5 percent in 2003, and has since declined again to 5.3 percent in 2008, nearly the level where it started in 1999.[39]

For the most part, this observation can be ascribed to the fact that fossil fuel consumption has continually declined in Germany since the introduction of the environmental tax reform (Figure 5.1). According to the German Federal Statistical Office, gasoline consumption in 2000 decreased by 4.5 percent compared to the previous year, and continued to decrease in 2001 and 2002 by 3.0 and 3.3 percent respectively, exceeding the previous average reduction of 2 percent due to general improvements in vehicle technology and transportation planning.[40] The targeted increase in energy costs has also created a definite incentive for behavioral change in other sectors, encouraging deployment of energy efficient technologies and processes, including alternative energy sources. Reductions of CO_2 emissions are estimated to have reached 3 percent annually, equivalent to 24 million metric tons of CO_2.[41] At the same time, revenues of the Ecological Tax Reform have been almost fully returned to taxpayers, with the largest share used for a gradual reduction of social security contributions. In 2003, for instance, roughly 16.1 billion euros raised through the tax reform were used to reduce and stabilize non-wage labor costs, allowing pension contributions to be lowered by 1.7 percent.[42] With hiring rendered less expensive, the Ecological Tax Reform is said to have promoted employment and thereby contributed to the creation of an estimated 250,000 new jobs.[43] A smaller fraction of revenues was used to subsidize the deployment of renewable energies and the modernization of buildings.

Unlike many of its European neighbors, Germany has emerged from the recession with a robust economy, thanks in large part to flourishing exports. At least part of this success is ascribed to Germany's dominant market share in various green technologies, a sector also employing a substantial share of the German workforce.[44] As a result, Germany has

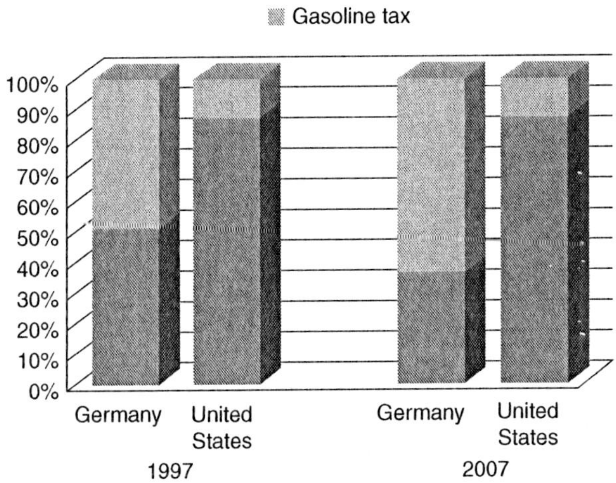

Figure 5.1 Share of taxes in gasoline price in 1997 and 2007
(*Source*: Author, based on data by the German Statistical Office and the Energy Information Administration).

earned wide recognition for its successful alignment of prosperity and sustainable growth. The greening of its economy is unmistakably also the product of targeted policy design and implementation, including the Ecological Tax Reform. As a former minister central to Germany's recent success has stated: 'green policy is merely good industrial policy'.[45]

Conclusion: Lessons from the German ecological tax reform

As they are anywhere, taxes are a politically sensitive issue in Germany. Unsurprisingly, opponents of the environmental tax reform – including the current ruling coalition – were quick to launch a determined media campaign against the proposed legislation. Given the complexities of its design, it was easy for critics to portray the tax reform as a mere increase in the fiscal burden, while downplaying or disputing the accompanying reduction in labor costs and expected employment benefits. Industry representatives have condemned the act for failing to simplify taxation rules, while worsening the competitive position of German industry with higher rates than in neighboring states. The Association of German Chambers of Industry and Commerce, for instance, decried the Energy Tax Act for being 'ineffective for purposes of energy policy, unfair in its distribution of the fiscal burden, and complicated in the procedure it sets out'.[46] At the other end of the spectrum, environmental advocates contended that the tax reform was a missed opportunity to eliminate subsidies for energy-intensive sectors and expand energy taxation to sectors traditionally exempt from fuel charges, such as aviation and shipping, within the narrow boundaries afforded by international and EU law.[47]

Despite such resistance from several sides, Germany's parliamentary system and its strict party discipline allowed the governing coalition at the time to pass the tax reform against partisan resistance. Ironically, the need to close a growing budget deficit has made the current conservative government, previously an ardent adversary of environmental taxes, now dependent on the revenue they yield. As the rationale and benefits of the tax reform have become more widely known, so has public acceptance of the incremental increases in energy cost. It stands to reason that better communication in the early stages of the tax reform could have alleviated some of the early concerns. Also, describing it as an environmental tax arguably did not help; more advisable might have been a stronger focus on the innovation and employment potential of the proposed tax. And clearly, a gradual and transparent trajectory of rate hikes was of central importance in making the tax reform acceptable in the first place. Ultimately, however, the positive outcome of the tax reform is the most compelling lesson from the German experience: contrary to the early fears, behavioral change and innovation prompted by the rise in energy prices have actually strengthened the German economy. Energy-efficient technologies are now among the fastest growing export products, and the incentive to reduce energy use has helped the German economy become more resilient to fluctuations in global oil and gas prices.

What the German Ecological Tax Reform has shown, therefore, is that environmental taxes, while no panacea, can be a useful addition to existing policies to promote greater energy efficiency and a transition to cleaner energy sources.[48] But that is not their only attraction. By shifting the tax burden from labor toward polluting actors, they also afford a less distortionary approach to raising revenues.[49] This benefits the labor market and at the same time provides impetus to certain industries. In Germany, environmental taxes have thus provided a tool in the ongoing effort to address some of the larger difficulties faced by modern-day economies, foremost the demographic trend apparent in a rapidly aging society, while simultaneously encouraging environmentally sound, innovative technologies. Given the robust pace of the German economy, fears that the tax reform would stifle growth and prosperity clearly seem overstated in hindsight. As the German Federal Environmental Agency rightly concluded, the Ecological Tax Reform delivered on its promise of improved labor conditions and greater sustainability, resulting in what it describes – in a typically German understatement – as a 'positive macroeconomic balance'.[50]

Notes

1. Federal Ministry for the Environment, Nature Conservation and Nuclear Safety (BMU) and Federal Ministry for Economic Affairs (BMWi), 'Energy Concept for an Environmentally Sound, Reliable and Affordable Energy Supply', available at www.bmu.de/files/english/pdf/application/pdf/energiekonzept_bundesregierung_en.pdf, accessed 15 March 2011.
2. Under the Energy Concept, primary energy consumption is to fall by 20 percent compared to 2008 levels by 2020, and at least 50 percent by 2050; renewable energy is to account for 18 percent of final energy consumption in 2020, and at least 80 percent of electricity consumption in 2050; and greenhouse gas emissions are to see cuts of 40 percent by 2020 and at least 80 percent by 2050, both relative to 1990 levels, see BMU et al., *supra*, note 1. For comparison, in the United States, the administration under President Barack H. Obama has pledged a cut in greenhouse gas emissions of 17 percent by 2020 relative to 2005 levels, which is roughly equivalent to stabilizing emissions at 1990 levels.
3. Among economists, of course, the notion of controlling externalities by imposing a system of unit taxes or subsidies has been discussed ever since it was proposed by Arthur C. Pigou (1920) *The Economics of Welfare* (London: Macmillan).
4. See inter alia Kalle Määttä and Michael Mehling (2010) 'Energy, the environment, and price-based instruments: environmental taxes as a tool for energy sustainability and climate policy', in Michael Rodi (ed.) *Realising the Paradigm Shift towards Energy Sustainability: Climate Change, Technological Innovation, and the Challenge of an Optimal Instrument Mix* (Berlin: Lexxion), pp. 49–64.
5. Official Records of Parliament (*BT-Drucks.*) 14/40 of 17 November 1998.
6. Official Records of the Federal Council (*BR-Drucks.*) 105/99 of 19 March 1999.
7. *Gesetz zum Einstieg in die ökologische Steuerreform* (hereinafter referred to as the Ecological Tax-Reform Act) of 24 March 1999, Federal Law Gazette (*BGBl.*) Part I (1999), pp. 378–384.
8. On the reasoning of the federal government, cf., inter alia, Official Records of Parliament 14/40 of 17 November 1998, p. 1 ff.; *cf.* also the Finance Committee, in: Official Records of Parliament 14/440 of 1 March 1999, p. 1, and section 3(3) of the Coalition Agreement between the Social Democratic Party (SPD) and the Green Party (*Bündnis 90/Die Grünen*) of 20 October 1998.
9. Starting on 1 January 1999, the euro replaced the deutschmark as the official currency for all financial transactions, initially as an accounting currency, from 2002 onwards also in general circulation.
10. Official Records of Parliament 14/40 of 17 November 1998, p. 2.
11. German Stability Pact of 1 January 1999, p. 14.
12. One-fifth of the normal rate, i.e. 0.4 pfennigs per kilowatt-hour on electricity.
13. The abatement, a 50 percent rebate off the normal rate, also applied to thermal storage heating stoves, albeit for social reasons.

14. At the time, the EU Competition Commissioner Karel van Miert expressed concerns that such a measure could violate EU state aid rules; the amended Ecological Tax Reform Act was eventually approved by the EU Commission on 21 April 1999.
15. This was also intended by the legislator, see Official Records of Parliament 14/40 of 17 November 1998, p. 10.
16. When used as vehicle fuel, thus, liquefied gas was only taxed at 25.57 pfennigs per kilogram and natural gas at 1.98 pfennigs per kilowatt-hour.
17. *Entwurf eines Gesetzes zur Fortführung der ökologischen Steuerreform* of 25 August 1999, in: Official Records of Parliament 14/1524 of 2 September 1999; 14/1668 of 29 September 1999.
18. See the minutes of the parliamentary proceedings, in Plenary Protocol 14/69 of 11 November 1999, pp. 6184–6215.
19. *Entwurf eines Gesetzes über eine ökologisch wirklich wirksame Umstellung der Besteuerung ohne Mehrbelastug für Bürger und Wirtschaft*, in: Official Records of Parliament 14/339 of 23 February 1999.
20. Plenary Protocol 14/69 of 11 November 1999, pp. 6184–6215, at p. 6211.
21. Ibid., at p. 6205 *et seq*. and p. 6208 *et seq*.
22. *Gesetz zur Fortführung der ökologischen Steuerreform* (hereinafter referred to as the Ecological Tax Reform Continuation Act) of 16 December 1999, in: Federal Law Gazette Part I (1999), pp. 2433–2440.
23. Plenary Protocol 14/69 of 11 November 1999; Official Records of the Federal Council 638/99 of 26 November 1999.
24. Official Records of Parliament 14/1524 of 2 September 1999, p. 9.
25. Reinhard Schultz, parliamentary speaker of the Social Democratic Party, calculated a reduction of labor costs of 115 billion marks over 5 years, in: Plenary Protocol 14/69 of 11 November 1999, pp. 6184–6215, at p. 6184.
26. Para. 1 Federal Act on Social Insurance Contributions for the year 2000 (*Beitragssatzgesetz 2000*) of 12 December 1999, in: Federal Law Gazette Part I (1999), pp. 2534–2551, at p. 2544.
27. See, for instance, Reinhard Schultz, in: Plenary Protocol 14/69 of 11 November 1999, pp. 6184–6215, at p. 6185.
28. With the exception of hydroelectric power plants, whose wattage may not exceed 10 megawatts.
29. Part B of the Reasoning for the Draft Federal Act on the Continuation of the Ecological Tax Reform (*supra*, footnote 17).
30. The tax rate on diesel used for public transportation, for instance, was increased by only 50 percent.
31. See Reinhard Loske, parliamentary speaker of *Bündnis 90/Die Grünen*, in: Plenary Protocol 14/69 of 11 November 1999, pp. 6184–6215, at p. 6191.
32. *Gesetz zur Neuregelung der Besteuerung von Energieerzeugnissen und zur Änderung des Stromsteuergesetzes* of 15 July 2006, Federal Law Gazette (*BGBl.*) Part I (2006), p. 1534 *et seq*.
33. See *Gemeinsam für Deutschland: Mit Mut und Menschlichkeit – Koalitionsvertrag von CDU, CSU und SPD*, 11 November 2005, p. 53.
34. See Council Directive 2003/96/EC of 27 October 2003 restructuring the Community framework for the taxation of energy products and electricity, Official Journal L 283, p. 51 *et seq*.

35. European Court of Justice (ECJ), Case C-240/01, *Commission of the European Communities v. Federal Republic of Germany*, European Court Reports (2004), p. I-04733 *et seq.*
36. *Verordnung zur Durchführung energiesteuerrechtlicher Regelungen und zur Änderung der Stromsteuer-Durchführungsverordnung* of 31 July 2006, Federal Law Gazette (*BGBl.*) Part I (2006), p. 1753 *et seq.*
37. See, inter alia, the critical remarks of Hans-Olaf Henkel, president of the Federation of German Industries (BDI), in *Die Welt*, 22 May 1999, p. 17; Karl Heinz Däke, president of the Confederation of Tax Payers (BdSt), in *Berliner Morgenpost*, 1 April 1999, p. 1; and the demands for a 'cessation of all eco tax plans' expressed by Hans Peter Stihl, president of the German Chamber of Commerce and Industry (DIHT), in *Frankfurter Allgemeine Zeitung*, 7 April 1998, p. 17.
38. Recommendations by the Committee, in: Official Records of the Federal Council 638/1/99 of 16 November 1999.
39. Damian Ludewig, Bettina Meyer and Kai Schlegelmilch (2010) *Greening the Budget: Pricing Carbon and Cutting Energy Subsidies to Reduce the Financial Deficit in Germany* (Washington DC: Heinrich Böll Stiftung), p. 15.
40. Ibid. In the meantime, gasoline consumption in the United States steadily rose, from 161,411 million gallons in 1999 to 167,730 in 2002.
41. Michael Kohlhaas (2005) *Gesamtwirtschaftliche Effekte der ökologischen Steuerreform* (Berlin: DIW), p. 14.
42. Markus Knigge and Benjamin Görlach (2005) *Die Ökologische Steuerreform – Auswirkungen auf Umwelt, Beschäftigung und Innovation* (Berlin: Ecologic), p. 5.
43. Kohlhaas, *Gesamtwirtschaftliche Effekte der ökologischen Steuerreform.*
44. Torsten Henzelmann (2010) 'Weltmarktführer beim Umweltschutz', *Harvard Business Manager* 30 (12), pp. 44–49.
45. Citing former environment minister Sigmar Gabriel: Stefan Theil, 'No Country Is More 'Green By Design', *Newsweek* Special Report, 7–14 July 2008. See also the analysis of Peter Debaere, in this volume, who discusses whether stricter environmental regulations can promote the emergence of competitive industries better able to succeed in the global marketplace.
46. DIHK, letter dated 12 May 2006, available at www.ihk-koeln.de/Navigation/FairplayRechtUndSteuern/Steuern/Anlagen/EnergieStGesStellungnDIHK.pdf, accessed 15 March 2011; translation by the author.
47. See, for instance, Grüne Liga e.V., letter dated 17 February 2006, available at www.grueneliga.de/aktuell/stellungnahme_biogene_kraftstoffe.pd, accessed 15 March 2011.
48. Terry Barker (1997), 'Taxing pollution instead of jobs', in Timothy O'Riordan (ed.) *Ecotaxation* (London: Earthscan), p. 163 ff., at p. 196; Anselm Görres, Henner Ehringhaus and Ernst Ulrich von Weizsäcker (1994) *Der Weg zur ökologischen Steuerreform* (Munich: Olzog-Verlag), p. 27 ff.
49. So-called double dividend of environmental taxes, see Daniel McCoy (1997), 'Reflections on the double dividend debate', in Timothy O'Riordan (ed.) *Ecotaxation* (London: Earthscan), p. 201 *et seq.*, at p. 201.
50. Umweltbundesamt (2004) *Quantifizierung der Effekte der Ökologischen Steuerreform auf Umwelt, Beschäftigung und Innovation* (Berlin: UBA), p. 2.

6
'Nuclear Power? No, Thank You!': Germany's Energy Revolution Post-Fukushima

Manuela Achilles

When on 11 March 2011 the 9.0 Great East Japan Earthquake and resultant tsunami crippled the cooling systems at the Fukushima Daiichi nuclear power station, thus sending several reactors into core meltdown, the German outcry against nuclear energy was almost unanimous. On Saturday, March 12, some 60,000 people demonstrated against the continued operation of one of the country's oldest nuclear power stations by forming a 45-kilometer human chain from the power plant to the regional capital. Two days later, more than 100,000 demonstrators took to the streets in 400 towns and cities across the nation. Faced with the rising public pressure, the federal government shut down the country's seven oldest reactors and imposed a technical audit on all nuclear power plants. In addition, the cabinet appointed an independent ethics committee on the safety of the nation's energy future.[1] Upon completion of the review process, Chancellor Angela Merkel announced on May 30 that Germany would phase-out nuclear power by 2022. Parliament passed the respective bill with an overwhelming majority on 8 July 2011. A heated public debate had come to an end: nuclear power was done for and over with.

Notably, this radical backlash against nuclear power did not occur in Japan but in a country over 5,500 miles (9,000 kilometers) removed from the Fukushima accident and arguably not directly affected by the radioactive fallout. The immediate formation of an inter-generational, cross-party consensus against nuclear energy is stunning, especially if one considers the economic risks of the phase-out decision. As this chapter is being written, Germany is the world's fourth largest economy and second largest industrial exporter, with 1 million jobs in energy-intensive industries such as aluminum and steel. National prosperity relies, among other things, on the steady availability of cheap energy. In abandoning nuclear power, Germany has committed to replacing one quarter of its electricity supply within a decade. How is the country to achieve this task?

The easiest way to close the looming power gap would perhaps be to increase reliance on fossil fuels or buy electricity from France, which generates three-fourths of its electricity from nuclear sources and has significant excess capacity. German popular opinion, however, rejects these options. Although natural gas has gained currency as a 'bridging technology', the aim is to phase out nuclear power and at the same time meet the country's ambitious domestic climate protection targets. The latter involves cutting carbon emissions by 40 percent by 2020, 55 percent by 2030, 70 percent by 2040, and 80–95 percent by 2050, relative to 1990 levels. To attain these targets, Germany needs to double the amount of electricity generated from renewable sources from 17 percent in 2011 to 35 percent by 2020. In addition, electricity consumption is to be reduced by 10 percent. The ultimate objective is to obtain 80 percent of all energy from renewable sources by 2050.[2] Requiring large investments in alternative technologies, as well as in new power lines and energy storage systems, this radical transformation of the German energy sector amounts to nothing less than a veritable 'energy revolution' and is increasingly described in these terms.

What drives this remarkable experiment in national self-transformation? Surely, the images of exploding reactors and displaced families televised around the world from the site of the Fukushima disaster triggered strong responses also in other countries. Germany, however, is the first major industrialized power to completely cease using a form of energy generation that its supporters describe as carbon-neutral, plentiful and cheap. Why is the highly developed nation abandoning a technology that continues to enjoy some level of support in the United States, France and even Japan? What accounts for the intensity of the German reaction when public excitement never flared up or quickly receded elsewhere? This chapter explores the German nuclear phase-out in two steps. The first part discusses the idea of a global nuclear renaissance promoted by industry circles as a weapon in the fight for energy independence and against climate change. Against this background, the German position appears particularly striking. To fully explain the rationale behind the country's shift away from nuclear and fossil fuels towards an untested renewable energy future, the second part looks towards the realms of history, politics and culture. As will be argued, the collapse of support for nuclear power in Germany is the outcome of fundamental societal changes that point back to the world historical conjuncture of 1945. The German case thus opens a window onto a discussion of sustainability that is sensitive to the global connectedness of environmental questions *and* the cultural-political determinants of the national response. The underlying question is whether Germany's energy revolution can be a template for other nations, including the United States.

Global perspectives

While the German phase-out policy condemns the country's commercial nuclear power program to a slow death, the industry's global future is more difficult to predict. As of December 2011, a total of 30 countries operated 434 commercial fission reactors with an installed capacity of about 370 GW (gigawatts or thousand megawatts).[3] These plants produced about 13 percent of the world's electricity and about 5 percent of the total primary energy supply.[4] Three-fourths of the global nuclear power generation was concentrated in eight countries. The United States had the largest nuclear fleet with 104 reactors, followed by France (58), Japan (50), Russia (33), South Korea (21), India (20), the United Kingdom (18) and Canada (18).[5] The majority of these facilities was over 20 years old and one quarter was over 30 years old.[6] The aging of the world's nuclear fleet has raised questions about the extension of reactor life spans from 40 to 60 years. These safety concerns have become more urgent after the destruction of the Fukushima Daiichi Nuclear Power Plant, which was first commissioned in the early 1970s.

The global nuclear fleet peaked at 444 reactors in 2002. Since then, permanent plant shutdowns have outnumbered new grid connections.[7] Increases in production capacities in industrialized countries were due mostly to technical upgrades that boosted the maximum heat output or power level of already existing reactors.[8] In 2008, no new reactor unit was started up for the first time since the beginning of the commercial nuclear age in the mid-1950s.[9] That year, however, saw 10 construction starts, followed by 12 in 2009 and 16 in 2010.[10] More than 65 reactors were under construction in 2011 (some of them, however, already for decades).[11] The building time world record is held by the Tennessee Watts Bar Nuclear Plant Unit 2 in the United States, which received its original construction license in 1973 and is scheduled to enter commercial production in 2013.[12]

The Fukushima effect

The Fukushima disaster has had a profound effect on global public opinion. A survey by the leading British polling company, Ipsos MORI, conducted in April 2011 reported that 62 percent of citizens in 24 countries across the globe opposed the use of nuclear energy, with a quarter of those having changed their minds after the Fukushima accident.[13] The most anti-nuclear nations were Italy, Germany, and Mexico at about 80 percent against. A multi-country poll for BBC World Service by the international polling firm, GlobeScan, found relatively high levels of public approval only in the United States, the United Kingdom, China, India and Pakistan. However, support for building new plants was a minority view even in those countries, hovering around 40 percent.[14]

Public opinion plays an important role in Western-style democracies and is considered an important factor in nuclear policy making.[15] This is true irrespective of the fact that representative governments are generally thought to have a mandate to deliberate policy decisions with expert bodies and industry circles. As discussed in more detail in part two of this chapter, the German political system was sensitive to anti-nuclear public pressure not least because of the strong structural and historical position of the environmentalist Green Party. In Italy, a popular referendum held in the wake of Fukushima thwarted Prime Minister Silvio Berlusconi's attempt to restart a nuclear program, which had been shut down in response to the 1986 Chernobyl accident.[16] Belgium, which drew more than half of its electricity from nuclear power in 2011, decided to stand by a 2003 law that limits the life-span of the country's seven reactors to 40 years and prohibits the building of new nuclear power plants.[17] Switzerland, which generated about 40 percent of its electricity from nuclear fission reactors in 2011, decided to phase out nuclear energy by 2034.[18]

In Japan, which derived about 30 percent of its electricity from nuclear power and had the third-largest global nuclear energy capacity in 2010, the industry's future is subject to considerable uncertainty. Public opposition to building new nuclear power stations has grown significantly after Fukushima.[19] Faced with its inability to contain the crisis, the national government has revised its pre-Fukushima goal of drawing about half of the country's electricity from nuclear power by 2030. At the local level, towns and prefectures have been refusing to allow nuclear reactors to be restarted after regularly scheduled routine maintenance.[20] The shutdown of virtually all Japanese nuclear power stations one year into the Fukushima disaster is not likely to be the final word, however. Since many rural communities depend on the industry for economic reasons, the technology is expected to remain part of Japan's energy mix for some time to come.

The case of the United States, which pioneered both the development of nuclear power and the formation of a vocal anti-nuclear protest movement,[21] highlights the multilayered interplay of political and economic factors in the nuclear energy field.[22] Two decades of rapid expansion culminated in renewed enthusiasm for the technology in the wake of the 1973 oil crisis, which put an end to decades of cheap energy. Once hailed as an energy source too cheap to meter, nuclear energy now signified independence from an energy market that saw oil prices quadruple in the course of just a few months. The global recession that followed the OPEC oil embargo, however, led to cuts in the predicted growth of electricity demand, thus discouraging investment in the capital-intensive technology. The nuclear core meltdown at the Three Mile Island nuclear plant in Pennsylvania in 1979 then caused the US Nuclear Regulatory Commission (NRC) to tighten its oversight. Resultant changes in the nuclear industry enhanced plant safety but also increased lead times and constructions costs, thus all but eliminating nuclear energy's profitability expectations. By the time of the Chernobyl accident in 1986, the industry was in decline mostly for economic reasons. In fact, all of the 104 reactors currently in operation in the United States were ordered prior to 1974, and less than half of the reactors on order in 1974 were ever completed.[23]

The US nuclear sector regained momentum at the beginning of the new millennium, when Republican president George W. Bush took up nuclear power as a clean alternative to global warming. Having rejected the ratification of the Kyoto Protocol in 2001, the Bush administration expedited the renewal process for existing nuclear plants and provided loan guarantees as well as tax incentives for the nuclear power industry. Licensing applications increased to a total of 28 by the first quarter of 2011,[24] but this temporary spike in orders fizzled in the wake of the late-2000s global financial crisis. Imploding gas prices and the failure of Congress to pass a federal carbon tax also played a role in diminishing the nuclear industry's growth potential.

In the absence of significant new construction, the US nuclear fleet is likely to gradually age out of existence. Out of a total of 104 reactors originally designed to last for 40 years, more than half had their licenses extended by two decades. Another 18 are under Nuclear Regulatory Commission review. In February 2012, the NRC granted a license to build and operate two reactors at the Vogtle Nuclear Power Plant in Georgia. The $14 billion project has been praised as a crucial threshold for an industry that has not had a new construction start in more than three decades. Most utilities remain apprehensive of the delays and cost overruns that accompanied the last waves of US reactor building in the 1970s and 1980s.[25] Skepticism is especially high in federal states with open energy markets, where their utilities cannot simply transfer costs to the consumer.[26] Legal challenges from anti-nuclear groups, as well as the evolution of both natural gas prices and federal emission rules, will also affect the prospects of nuclear energy in competition-driven, liberalized market economies such as the United States.

In France, which has the highest per-capita investment in nuclear energy in the world, the general political climate became more critical of the technology in the wake of Fukushima.[27] According to a poll by the French Institute of Public Opinion published in June 2011, 62 percent of French people wanted a progressive halt to the country's nuclear power program in the next 25 to 30 years.[28] A survey conducted by the French Institute for Radioprotection and Nuclear Safety in Fall 2011 indicated that 55 percent of the population rated the risks from nuclear power plants high, and only 24 percent trusted the authorities to protect the public against this danger.[29] It is questionable, however, whether France will significantly reduce its investment in nuclear power. After all, the government-owned utility EDF (*Electricité de France SA*) produces most of its electricity from nuclear power. The state-owned nuclear conglomerate Areva, on the other hand, is a key driver of jobs and exports, with revenues of $12.3 billion and 48,000 employees in 2011.[30]

As France struggles with the question of whether to maintain their nuclear investment or re-channel resources into (the search for) renewable alternatives, Areva is looking for new markets abroad. Growth potential is seen in Eastern Europe, the Middle and Far East, South Africa, India and especially in China, which alone accounted for more than half of the new construction starts in 2010.[31] Having initialized its commercial nuclear program only in 1985, China had 15 reactors in operation and 26 under construction in 2011, with 51 more planned and 120 proposed. The government's objective is to have 43 GW of reactors in operation by the end of 2015 and 80 GW by 2020.[32] This would amount to the second largest installed nuclear capacity behind the United States.[33] Since the Chinese government ties domestic market access to technology transfers, nuclear exports from France, Canada, the United States and Russia are preparing the ground for the country to become a large-scale exporter of nuclear technology in the future. Popular opposition against siting new plants near population centers is growing in China, however, and new safety regulations drafted by the government in response to the nuclear accident at Daiichi have yet to be revealed.[34]

It appears, then, that the Fukushima accident has had an uneven effect if considered on the global scale. While some industrialized nations reduced or halted their nuclear programs, countries with fast-growing economies and rapidly rising electricity demand continue to press for nuclear fleets of their own. The anticipated migration of nuclear construction further 'south' to low-income and low-middle income economies such as Bangladesh and Vietnam has raised the specter of a global divide over the safety of nuclear energy, separating the world's wealthiest and poorest nations. Driven by aggressive marketing competition between the mostly Western exporters of nuclear technology, this development could result in a nuclear market at two speeds: a high-tech, high-safety mode for developed countries and a lower safety mode for emerging countries.[35] In addition, the overlap of the new markets for nuclear power with political hot spots in the Middle and Far East has increased concerns over the danger of nuclear weapons proliferation. The international tension sparked by Iran's nuclear power program is a case in point. Rather than deciding the controversy over the industry's global future, the Fukushima accident has thus stirred up and complicated the nuclear energy debate. To understand Germany's decision to abandon the technology, it is necessary to understand the global context and to identify the determining factors specific to Germany's case.

A clean template for the world?

Germany is a highly developed and densely populated country in the center of Europe. Shaped by the historical legacies of two world wars and the Holocaust, the population has become acutely attuned to the international context to which it is bound. After the collapse of the Nazi Third Reich in 1945, the defeated nation was divided along the major fault lines of the Cold War. The German Democratic Republic (GDR) became tied to the Warsaw Pact and the Council for Mutual Economic Assistance (COMECON), and boasted the highest per-capita production output in the Eastern bloc. The Federal Republic of Germany (FRG) integrated into the European Community and the North Atlantic Treaty Organization (NATO) and became one of the most developed economies in the world. West Germany, which was wealthier and more used to a process of democratic decision-making, dominated the unification process in the wake of the fall of the Berlin Wall in November 1989. Although the GDR legacy retains a strong influence in the country's east, it is fair to say that the unified nation stands in the (legal) tradition of the Federal Republic.

Germany's post-unification economy is export-driven and supports a high standard of living. Being able to take their physical safety and economic security for granted, large segments of the population place a high price on post-materialist values such as personal freedom and environmental concerns. The prosperous nation is poor in raw materials, however, and derives most of its considerable energy supply from imported sources. In fact, Germany is one of the world's major consumers of energy and the largest electricity market in Europe. In 2010, more than half of the gross electric power generation of 621 billion kWh was based on lignite (23.7 percent), hard coal (18.7 percent) and natural gas (13.6 percent). Nuclear energy had a share of 22.6 percent. Renewables (wind, water, bio-mass, photovoltaic) accounted for 16.5 percent.[36] While this distribution reflects a heavy reliance on fossil-fuel-fired power generation, the cost-competitiveness of coal and natural gas is expected to decline with increasing carbon pricing through the European Union's greenhouse gas emission allowance trading scheme. Germany's existing nuclear power plants, on the other hand, are lucrative because the investment expenses of most reactors have been amortized and the fuel is not subject to carbon-based energy taxes. According to some estimates, a large plant with a capacity of 1,300 megawatts generates a profit of at least €1 million ($1.3 million) per day, and smaller, 800 MW plants, about half as much.[37]

The German nuclear phase-out goes far beyond the profit interests of a particular energy sector, however, as the accelerated energy switch is bound to have profound implications for the entire society. Germans will have to dramatically increase the energy efficiency of their buildings and machinery. They will have to develop energy storage technologies and build massive high-voltage power lines to support a smart grid. Large wind farms and transmission lines will cause dramatic transformations of the environment, thus raising questions about the compatibility of the energy revolution with the green ideas that inspired it in the first place. Last but not least, the energy transition will be expensive. According to the *World Nuclear News*, Germany will have to invest about €25 billion ($33 billion) per year until 2020 to achieve its post-Fukushima energy targets.[38]

A second cluster of concerns revolves around the uncertain consequences of the nuclear exit for the price and supply of electricity. Critics of the nuclear phase out argue that the accelerated transition to renewables is neither practical nor economically feasible. The major question is whether the new technologies can deliver the steady flow of cheap energy provided by traditional base-load power sources such coal-fired and nuclear plants. Germany's 'big four' electricity producers, which are the major shareholders of the country's nuclear power stations, are projecting a heightened probability of power outages especially in winter when high demand meets a reduced supply of wind and solar power.[39] Large manufacturers are warning that the nuclear exit will increase pollution and electricity prices, making German production less competitive on the global market and undermining the country's leading role on climate policy. The chief executive of the German energy giant RWE, which distributes electricity to more than 20 million customers, described the anticipated transition to renewable sources as a 'Herculean task'.[40] The *Financial Times* labeled Germany's energy transformation 'one of the biggest gambles ever made by an advanced industrial country on renewable energy'.[41]

Faced with the vocal criticism of a business community that usually counts amongst its staunchest supporters, the Merkel administration has been steadfast in noting that the people want the transition to renewable energy and are prepared to pay for it.[42] In addition to invoking its democratic mandate, the center-right government seeks to focus the debate on the added economic value of green technologies. According to Environment Minister Norbert Röttgen of the Christian Democratic Union, investment into renewable sources of energy makes sense because it will create the high-tech employment opportunities of the future. The German Ministry of the Environment (BMU) reported in July 2011 that the number of jobs connected with renewable energy more than doubled by 2010 to 367,400 from 160,500 in 2004 and could be expected to rise to over half a million jobs in 2030.[43] The ministry promoted these numbers together with a broad information campaign, invoking the future as being 'made in Germany'. Röttgen summarized the official policy line when he described the German people as trailblazers on the path to a new energy era.[44] 'If we proceed successfully', the minister asserted, 'others will follow'.[45] Stephan Kohler, the chief executive of the semi-public German Energy Agency (*Deutsche Energie-Agentur*, DENA), reiterated this claim by suggesting that Germany could become a high-tech, clean-energy model for the world.[46]

The German desire to become pioneers on the 'path into a sustainable modernity' is remarkable, considering that the country belonged to the laggards when modern environmental policies first began to be developed in the 1960s, not least in the United States.[47] As we have seen, nuclear power has since gained status as a zero emissions industry. Why does the technology appear as part of the problem and not of the solution in Germany? The country could just as well apply its widely respected engineering know-how to the improvement of nuclear energy generation. A comparison with neighboring France throws the German situation into sharp relief. Both countries are highly developed but resource-poor EU member states that need to limit carbon dependence while at the same time ensuring energy security and market efficiency. The commercial use of nuclear power for energy generation should make as much (or as little) economic sense in Germany as in France.

Notably, the energy profiles of France and Germany were quite similar around the time of the 1973 oil crisis. Revealing a heavy reliance on imported fossil fuels, the oil shock triggered an urgent search for alternatives in both countries. The French government's massive drive to increase the nation's energy security resulted in the nuclear construction boom of the following decades. The nuclear sector also expanded in Germany, albeit to a lesser extent. The divide grew more explicit in the 1980s, when France turned to nuclear power not only as a means to gain energy independence, but to also mitigate aerial pollution from coal-fired power plants. With the adoption of the 1997 Kyoto Protocol and the launching of the European Union Emission Trading Scheme in 2005, the French government pursued nuclear energy as the primary means through which to reach the country's greenhouse emissions targets. Germany faced the same challenges but progressively addressed them differently. As nuclear energy moved into the very core of the French energy policy, the German nuclear energy expansion tapered off and finally stagnated at a quarter of the nation's energy supply. Electricity generation from renewable sources, on the other hand, gradually increased to reach a share of 20 percent in the first half of 2011. Figure 6.1 illustrates the share of nuclear energy in the total electricity of France and Germany from the early 1970s to the present. Since there is no obvious economic reason for the divergence in the French and German energy approaches, it is hard to understand the respective positions without reference to the political sensibilities rooted in history, politics and culture.

Nuclear power, democratic change and the legacies of history

It is important to note that at its introduction in the 1950s, the commercial use of nuclear power stood for technological innovation, economic growth and societal progress also in Germany. In fact, the German nuclear industry received large subsidies and tax benefits, as well as substantial R&D investments, far into the 1990s. And yet, the technology's positive appreciation was complicated by the fact that the nuclear breakthroughs of the 1940s had been driven by fears of a Nazi-built atomic bomb. Although Hitler's regime never acquired nuclear weapons capability and Germans eventually came to understand their defeat in World War II in

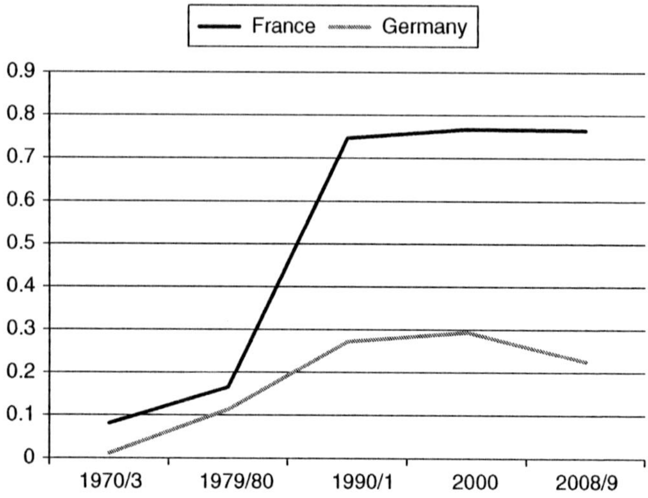

Figure 6.1 Share of nuclear energy in total electricity production: France vs Germany from 1970–2009

Note: The data for 1970 and 1980 focus on West Germany only.

Source: Country Nuclear Power Profiles, IAEA, Vienna, 2011.

terms of liberation, the country's legacy of military aggression required a strict separation of the technology's civilian and military uses. As France staked its claims to global leadership on its status as a fully equipped nuclear weapons state,[48] the demilitarized and divided Germany was forbidden to make or possess such arms and later affirmed the status quo by signing the Nuclear Non-Proliferation Treaty in 1975. Granted that the unified nation has the capacity to build all things nuclear and commands US nuclear warheads under NATO, German history still conveys a suspicion about the technology that its western neighbor does not share. France traces its proud involvement to Marie Curie's pioneering research on radioactivity. German nuclear science, on the other hand, is tainted by its compliance with Nazi politics.

Wary of Germany's military-industrial potential, the occupation authorities forbade any fission research in the defeated country in 1945. The build-up of nuclear science programs in the divided nation was an effect of the gradual political and economic integration of the respective territories into the cold war systems of alliances. Signaling the country's return into the circle of the world's leading industrialized nations, West Germany opened its first nuclear power plant in the Bavarian town of Kahl in 1960. Five years later, Communist East Germany started up the Rheinsberg Nuclear Power Station about 75 kilometers north of Berlin. The GDR eventually drew about one-tenth of its electricity supply from five nuclear reactors concentrated at the Ludmin Nuclear Power Plant near Greifswald on the Baltic Coast. Most of the country's electricity stemmed from lignite-fired thermal power stations. More consequential was the East German capacity for uranium mining in the Ore Mountains near the Czech border. Having produced a total of about 220,000 tons of uranium for the Soviet Union,[49] the mines were closed after German unification in 1990. East Germany's nuclear reactors were shut down for safety reasons. All remaining atomic power stations were built by the West German, Kraftwerk Union AG, a subsidiary of Munich-based engineering conglomerate, Siemens. The newest reactor unit in operation came online in 1989.

The expansion of the West German commercial nuclear power sector in the 1970s and 1980s overlapped with the formation and consolidation of the post-war peace movement, which reacted to the tensions of the cold war. Until German unification in 1990, nuclear missiles were installed in West Germany by the United States and in East Germany by the Soviet Union. The fact that the divided nation was a likely battlefield in a nuclear confrontation that it could not control explains the particular national inflection of the West German pacifist movement. Combining an intense fear of nuclear death with a decided renunciation of the country's history of military aggression and genocide, broad strata of the population rejected the rearmament of the divided nation in the 1950s.[50] The anti-military sentiment that first manifested itself in response to the stationing of nuclear weapons on German soil continues to resonate in the country's foreign policy today. Having marched to the battlefield twice to return to a materially and morally destroyed homeland, German public opinion no longer supports the rhetoric of honorable death for the fatherland and seeks to avoid military involvement in world affairs. In historical perspective, Germany's often lamented failure to assume a global military presence commensurate with its economic power is the result of a pacifist societal reorientation that bore first fruit in the 1950s.

The anti-nuclear pacifist protests of the immediate post-war era became more systematic in the 1960s when a new generation of students subjected the status quo to a comprehensive criticism that extended from the structural crisis of the universities to the exploitation of colonialized people, women and the environment. At the core of the protests stood the radical rejection of authoritarianism, imperialism and capitalism. The generational push for fundamental social and political change that erupted in many Western countries in the 1960s derived its specific historical dynamics in Germany from the rejection of a national consensus that imagined the German people as victims rather than perpetrators of Nazi crimes.[51] Gudrun Pausewang (born in 1928), the author of the two single most important anti-nuclear youth novels in Germany,[52] draws a clear link between her writing and repressed national guilt. Remembering the disgust she felt when her elders answered questions about their Nazi involvement with silence, Pausewang explained in a post-Fukushima talk show that she joined the anti-nuclear environmentalist movement because she wanted to speak out about the dangers she saw for her own and future generations. Silence was not an option, Pausewang noted; she wanted to undercut any future use of the standard post-war apology: 'We didn't know about it'.[53] There is ample evidence that this heightened sense of personal and historical responsibility was widespread among writers and activists that came of age at the end of World War II and in the 1960s. The critical questioning of the parents' refusal to admit to the organized murder of millions of people triggered a democratic opening of post-war German society that changed the country from the bottom up.

The West German anti-nuclear movement came into its own in the early 1970s, when an unusual alliance of local farmers and students from nearby Freiburg University occupied a nuclear construction site at Wyhl in the wine-growing region of southwest Germany. Media coverage of police officers dragging away peacefully demonstrating citizens turned resistance against nuclear power into a national concern. The administrative court eventually withdrew the plant's construction license and the land became a nature reserve. The Wyhl experience encouraged the formation of similar citizen action groups near other planned nuclear sites, but success was not guaranteed. While a fast breeder reactor at Kalkar in North Rhine-Westphalia and the Bavarian nuclear fuel reprocessing plant at Wackersdorf were aborted, other large constructions projects such as the controversial Brokdorf nuclear power plant on the North Sea coast west of Hamburg were completed and went online.

As the oil-shock-induced, nuclear construction boom subsided in the 1980s, the anti-nuclear protest movement turned to the reprocessing and storage of nuclear fuel. The demonstrations culminated in Gorleben, a municipality of about 700 inhabitants in the far northeast corner of Lower Saxony. Situated on the left bank of the Elbe river in a region also known as the Wendland, this small town was to accommodate the world's largest reprocessing plant together with a deep geological repository for highly radioactive waste. Not unlike the Wyhl experience, resistance against the project forged an uneasy alliance of local citizens and the urban Left.[54] Whereas anti-nuclear city dwellers carried out sit-down blockades or chained themselves to roads and railways, farmers blocked access with heavy machinery or by driving flocks of sheep onto streets. They also dispersed liquid manure at the exploratory drilling site and lined up spectacular tractor parades when taking their grievances to the local capital. The protests culminated in the founding of the 'Free Republic of Wendland' at the proposed site of the final repository. When the police finally evacuated the improvised 'peace village' after 30 days of existence in June 1980, the small encampment became a symbol of the collective alternative lifestyle that has since become green legend.

The provincial government eventually abandoned the Gorleben reprocessing plant as 'not politically feasible', but continued to pursue the exploration site for the final repository.[55] Shipments of spent fuel elements from reprocessing plants in France (La Hague) and Britain (Sellafield) to an intermediate storage facility for vitrified high-level nuclear waste kept Gorleben in the public eye. In November 2010, an unprecedented 50,000 people protested against the transport of eleven dry storage 'Castor' casks containing nuclear waste from the reprocessing of German spent fuel at La Hague. It took 17,000 police to secure the transfer from France to its Gorleben destination. The colorful presence of many youth at the Castor demonstrations testified to the fact that anti-nuclear protest culture had crossed over to the next generation. In present-day Germany, the movement's symbolism unites people of all ages and ranks.

From the fringes to the center: Green ideas go mainstream

An important station on the anti-nuclear movement's journey from the extra-parliamentary fringes of the 1960s into the mainstream of German society and politics was the foundation of the German Green Party (*Die Grünen*) in 1980.[56] Confronting a political elite they perceived as ossified and undemocratic, Green Party delegates entered the lower house of German parliament (*Bundestag*) in 1983. The environmentalist movement was also well developed in France at that time, but the German proportional voting system proved more favorable to the expression of minority opinions in government. Whereas the winner takes all in the French semi-presidential system, German elections hardly ever produce absolute majorities. Instead, the greater part of the vote tends to split between the conservative union (CDU/CSU) and the social democrats (SPD), with smaller parties sharing the minority vote. The number of political parties to be represented in parliament is limited by the requirement to win at least five percent of the vote. This constitutional qualification was introduced after World War II in response to the fragmentation of the Weimar political party system, which prevented the formation of stable democratic majorities in the early 1930s. The introduction of the 'five-percent clause' resulted in the formation of a three-party system that featured the business-friendly Liberals (FDP) alongside the CDU/CSU and SPD. Since grand coalitions between the two larger parties were the exception rather than the norm, the FDP assumed the role of 'king-maker'.

The arrival of the Greens on the national parliamentary stage added a new dimension to the political system of alliances, although the practical implications manifested themselves only gradually. Initially, there was little love lost between the self-declared 'green' outsiders and the 'black' (conservatives), 'red' (social democrats) and 'yellow' (liberal) Bundestag establishment. The tensions were lower in the regions, where political coalitions were more pragmatic and less ideological. As Green party members began to enter the provincial governments, another characteristic of the German political system came into play. Unlike unitary France, the German federal republic divides authority between the (federal) political center and 16 states or *Bundesländer*. While the latter act in part as executors of federal law, they also have considerable powers of their own. This is especially relevant in the areas of policing, education and taxation. Federal legislation that impacts regional interests requires the approval of the federal council (*Bundesrat*), which represents the provincial governments at the national level. Dissenting majorities in the upper and lower houses of parliament can thus have profound effects on national policy making.

Germany's federal structure plays an important role in the regulation and administration of the country's nuclear energy program, while the regional governments are the responsible licensing authorities for nuclear power plants and repositories. The Federal Office for Radiation Protection (BfS) is the licensing authority for interim storage sites and transport and the operator of a final repository.[57] The Federal Ministry for the Environment (BMU) supervises the BfS and states and has power to issue guidelines and directives in cooperation with other responsible agencies at the national and regional level.

The legal basis for the construction and operation of Germany's nuclear plants is the Atomic Energy Act (AtG).[58] Designed in the late 1950s to regulate the industry's expansion, the AtG was amended in 2002 and 2011 to write Germany's nuclear exit into law. In its most recent formulation, the act's purpose is 'to phase out the use of nuclear energy for the commercial generation of electricity in a structured manner, and to ensure on-going operation up until the date of discontinuation'.[59] For the duration of the nuclear phase-out, the continued operation of Germany's nuclear power stations remains a shared responsibility of the federal government and states.

The characteristic interlocking of federal and regional competencies in the regulation and administration of nuclear power had profound consequences for the ascent of the anti-nuclear movement into government. Although the Greens lacked a potential coalition partner at the national level, their participation in provincial governments provided them with both regional administrative power and legislative seats in the federal council. The gradual promotion of Green Party members, such as Joschka Fischer or Petra Kelly, into positions of political prominence and administrative responsibility rendered German political culture more open and flexible. At the same time, the Greens' adaptation to the political system brought its delegates into conflict with the radical groups and ideas that had founded the party. This problem was especially pressing when provincial green ministers oversaw the operation of nuclear plants rather than closing them, or mobilized the police against violent anti-nuclear protesters.

The loss of political support at the radical fringes was offset by gains amongst the new middle classes. As the 1968ers entered the professions and took charge of the bureaucratic institutions they had previously criticized if not rejected, their distrust of the state and capitalism gave way to a participatory spirit that created jobs and paid taxes. Serving as an integrative ideological bond, the nuclear exit remained non-negotiable for the Greens. As no other political party supported this position on principle, the Greens remained isolated at the federal level.

The situation changed when Germans confronted nuclear fallout from radioactive clouds released by the Chernobyl nuclear disaster on 26 April 1986. Concerns about the contamination of the environment, and especially of produce and other foods, deepened the anti-nuclear opposition and at the same time led to a broadening of the environmental powers of the German state. Indeed, it is no coincidence that the Federal Environment Ministry (BMU) was founded on 6 June 1986 in the direct aftermath of the Chernobyl accident. The institutional concentration of environmental powers at the highest government level corresponded to the perceived severity of the crisis as well as to the delayed and disorganized nature of the federal response. As the center struggled to determine safety levels for meat, milk and vegetables, the state of Hesse issued recommendations more frequently and that were significantly stricter than in any other state.[60] Hesse's openly deviationist course originated with a provincial government that had given the environment ministry to the Green Party. The founding of the BMU was to both strengthen the center's position vis-à-vis green pressures from the regions and improve the national disaster response. In turn, the concentration of responsibilities at the federal level was instrumental in elevating (regionalized) environmental policy into a national concern that had to be addressed at the highest level of German politics.

Notably, the new cabinet post originated with a center-right government that supported nuclear power despite sustained domestic protest against the industry's further expansion. Helmut Kohl, who served as Germany's chancellor between 1982 and 1998, oversaw the opening of several nuclear power stations and upheld his cabinet's pro-nuclear course after Chernobyl.[61] As in other industrialized nations including France and the United States, energy security provided a key argument in support of the technology. A second increasingly important justification revolved around the global mitigation of climate change. Indeed, much of the Kohl administration's efforts to reduce CO_2 emissions went into nuclear energy. Still, the research and development of alternative energy sources such as wind and solar also received increased funding, as well as infrastructural support.[62] Germany's pioneering electricity feed-in law, for instance, came into force during Helmut Kohl's third term in 1991. It required electricity utilities operating the public grid to ensure grid access and pay fixed prices (feed-in tariffs) for electricity generated from hydropower, wind energy, solar energy, landfill gas, sewage gas and biomass. The economic burden imposed by the law was borne by electricity suppliers and their customers; public funds were not used.

A comparative perspective highlights the particularities of the German energy path that began to emerge in the 1980s. As indicated in the global survey section of this article, the G.W. Bush administration also employed the energy security argument in support of the industry's expansion. In the United States, however, public concern with climate change mitigation was too weak to rationalize substantial investment in nuclear and/or renewable alternatives to fossil-fueled power generation. It was a combination of slackening electricity demand, increased regulatory oversight, rising nuclear construction costs and decreasing natural gas prices that determined the industry's fate in the United States.

In Europe, climate change mitigation became a deep political concern both domestically and at the highest levels of EU policy making. The relatively high public acceptance of nuclear power in France allowed for the state-driven expansion of the industry in the 1980s. The relative strength of the anti-nuclear opposition prevented a similar outcome in Germany. Having claimed global leadership in the climate change arena, however, the Kohl administration could not easily abandon the goal of reducing greenhouse gas emissions. Responding to the high public concern for green issues it had helped to create, the center-right government continued to push for climate change solutions at home and abroad. The successful completion of the Kyoto Protocol in 1997, for instance, has been credited in part to the preparatory work of Angela Merkel (CDU), then Minister of Environment, at the 1995 Conference of the Parties (COP1) conference in Berlin.[63] Consequently, it should be recognized that the German energy revolution is not simply a creation of the political Left. One trajectory can be traced to a conservative coalition that saw nuclear power as part of a low carbon energy mix and found the policy path blocked by the anti-nuclear movement. The pursuit of climate change mitigation, in the absence of a strong nuclear option, generated policy support for renewable alternatives in conservative circles as well.

Another historical trajectory leading from the 1980s to the post-Fukushima energy switch surfaced with the social democrats' renunciation of nuclear power in the aftermath of the Chernobyl accident. This was a serious change in direction for a party that had supported the technology unequivocally for more than three decades. Helmut Schmidt, Germany's Social Democratic Chancellor from 1974 to 1982, had pushed the program's expansion against both strong grassroots resistance and increasing disagreement within his own party. By the 1980s, however, the competition for green votes had all but eroded the social democrats' pro-nuclear stance. The internal opposition carried the day in August 1986, when the party congress at Nuremberg passed a resolution to abandon nuclear power within 10 years.[64]

The SPD's programmatic reversal on the issue of nuclear power prepared the ground for a coalition with the Greens by removing a major ideological obstacle. Twelve years later, the closely fought 1998 federal elections produced a parliament with a viable red-green majority. Gerhard Schröder, the Social Democratic governor of Lower Saxony, replaced long-serving Helmut Kohl in the chancellery. Joschka Fischer, completing his life journey from archetypal 1960s street activist to respectable political party leader, accepted the posts of vice-chancellor and foreign minister.[65] Anti-nuclear politics had finally arrived at the heart of German politics.

Although the election had been dominated by high unemployment and the need to reform the German social security system, the nuclear question remained high on the new coalition's agenda. Embarking on a drawn-out process of consensus finding with the country's big-four energy utilities, the Greens gave up their immediate exit strategy and accepted a gradual phase-out. Two years into its first term, the Schröder government announced a pragmatic compromise that limited reactor lifespans to 32 years while securing the operation of the existing plants. No further licenses for the construction and operation of new nuclear power plants would be issued. The agreement included a ban on the delivery of spent fuel elements for reprocessing and obligated nuclear power plant operators to set up intermediate onsite storage facilities for spent fuel. The German parliament approved the amendment to the 1959 Atomic Energy Act on 14 December 2001, and it was passed by the federal council on 1 February 2002.[66]

Complementing the nuclear phase-out, the Schröder administration expanded the feed-in law, first introduced a decade earlier under Helmut Kohl, into the more comprehensive Renewable Energy Law (*Erneuerbare Energien Gesetz*) in 2000.[67] The law has been designed to provide renewable energy generators with a reasonable return on their investment and is reviewed on a regular basis. Similar tariff regulations have been introduced in more than 50 countries at the national or provincial level, including Australia, India and the United States.

Schröder's red-green coalition won a second term in 2002, but political support was faltering over the government's tough labor and welfare reforms. Early elections in 2005 produced a 'grand' coalition between the SPD and CDU/CSU. Angela Merkel became the first female chancellor of Germany. Her administration's position on the nuclear issue was split. The Social Democrats, now in the position of the smaller government partner, continued to support the phase-out it had written into law. The Christian Democrats wanted to reopen negotiations with the industry, but found themselves bound to a coalition agreement that guaranteed the 2002 legislative amendments. It was only when the conservative union was in the position to form a governing coalition with the business-friendly Liberals after the 2009 federal elections that the nuclear exit stood on the agenda again. And yet, the political turn-around was cautious and came with a heavy price tag. A law written for the approval solely of the Bundestag extended the maximum operation time for Germany's 17 reactors by an average of 12 years, thus stretching the nuclear exit from 2021 to 2036. In return, the utilities agreed to a nuclear fuel tax to be used to subsidize the move away from nuclear energy and fossil fuels to renewable energy. Chancellor Angela Merkel took great pains to emphasize that the phase-out would continue, albeit at a slower speed. Nuclear power was to 'bridge' Germany's transition into a nuclear-free and low-carbon energy future.

The anti-nuclear opposition did not accept the official rationale, but rather aimed to revitalize a protest movement that had calmed down since the 2002 compromise with the nuclear industry. On 18 September 2010, tens of thousands of Germans surrounded Chancellor Merkel's office in demonstrations that organizers described as the biggest of their kind since the 1986 Chernobyl disaster.[68] But although the extension of reactor life spans was unpopular, the Merkel administration seems to have trusted its ability to control the negative political fallout. It is fair to assume that the government expected the anti-nuclear protests to die down, with a majority being willing to accept a prolonged exit as the pre-condition for the widely desired clean energy transition. The position of Federal Environment Minister Röttgen is a case in point. Having first strongly rejected the life span extension, the minister later reworked Chancellor Merkel's nuclear extension into a 'great success' on the road to the era of renewable energy.[69]

The Fukushima accident half a year later prevented a return to compromise conditions, tipping the political scales towards the anti-nuclear opposition. Faced with crucial provincial elections in Baden Württemberg and Rhineland-Palatinate on 27 March 2011, Chancellor Merkel almost immediately imposed a three-month moratorium on her own cabinet's extension of reactor life spans. Nuclear power stations that had started operation in 1980 or earlier were shut down. The new legislation to be worked out in the following months effectively returned to the nuclear exit strategy put in place by the Schröder government a decade earlier. This time, however, the anti-nuclear consensus included the conservative party, which makes the decision difficult to reverse.

The conservative party's struggle to regain the political initiative in the aftermath of the Fukushima nuclear disaster did not succeed in the short run. Riding on a wave of public support, the Greens surged in noted state elections at the expense of all other political parties. In the Rhineland-Palatinate, the ruling social democrats suffered large losses and were forced to form a coalition government with the Green Party. In Baden Württemberg, which is a traditional conservative stronghold and one of Germany's economic powerhouses, the Greens doubled their voter share and won their first-ever state governorship. The party's success in the first post-Fukushima elections marks an end as well as a new beginning for the Greens, who have lost a defining political issue and need to remake their political identity. At the same time, the resolution of the nuclear question has turned the environmentalist party into an attractive potential coalition partner for Chancellor Merkel's conservative union. As the conservatives become more environmentalist and the Greens more business-oriented, their eventual political coalition at the federal level would be congruent with the broad societal consensus against nuclear energy and in support of a renewable energy future in post-Fukushima Germany.

Notes

1. Ethics Commission for a Safe Energy Supply, 'Germany's energy transition – A collective project for the future', Berlin, 30 May 2011.
2. 'The path to the energy of the future – reliable, affordable and environmentally sound', BMU website, http://www.bmu.de/english/energy_efficiency/doc/47609.php, accessed 19 March 2012.
3. World Nuclear Industry Status Report 2010–2011, p. 12.
4. World Nuclear Industry Status Report 2010–2011, p. 11; IAEA Annual Report 2010, (www.iaea.org), p. 3.
5. Website of Nuclear Energy Institute (NEI): http://www.nei.org/newsandevents/newsreleases/inpo-compiles-timeline-of-fukushima-events-after-japan-earthquake-tsunami, accessed 2 December 2011.
6. IAEA Annual Report (2010), p. 6. More critical: The World Nuclear Industry Status Report 2010–2011, p. 8.
7. World Nuclear Industry Status Report 2010–2011, p. 12.
8. Cf. 'Power Uprates for Nuclear Plants.' Website of the United States Nuclear Regulatory Commission, http://www.nrc.gov/reading-rm/doc-collections/fact-sheets/power-uprates.html, accessed 20 February 2012. See also: World Nuclear Industry Status Report 2010–2011, p. 13.
9. World Nuclear Industry Status Report 2010–2011, p. 7.
10. IAEA 2010 report, p. 1; World Nuclear Industry Status Report 2010–2011, p. 10.
11. IAEA 2010 report, p. 10.
12. 'TWA to manage Watts Bar 2 project.' *World Nuclear News*, 20 October 2011.
13. Ipsos Global@dvisory, 'Sharp worldwide drop in support for nuclear energy as 26 percent of new opponents say Fukushima drove their decision', Ipsos.com, 20 June 2011; Damian Carrington, 'Citizens across world oppose nuclear power, poll finds', *The Guardian*, 23 June 2011.
14. 'Opposition to Nuclear Energy Grows: Global Poll', Online – International News Network, http://www.onlinenews.com.pk/details.php?id=186243. For a comparison of the IPSO Mori and GlobeScan polls, see Richard Black, 'Nuclear power "gets little public support worldwide"', *BBC mobile*, 24 November 2011.
15. Cf. Felix Kolb (2007) *Protests and Opportunities: The Political Outcomes of Social Movements* (Frankfurt am Main: Campus Verlag), p. 211 and ff.
16. Barry Moody, 'Italians say no to nuclear energy in referendum', *Reuters*, 13 June 2011.
17. Alessandro Torello,'Belgium Reviews Timing on Nuclear-Power-Exit', *The Wall Street Journal*, 15 February 2012.
18. 'Swiss to phase out nuclear power by 2034', *Swissinfo.ch*, 25 May 2011; Philippe Clavel 'Switzerland nuclear power phaseout approved by lawmakers', *Huffington Post*, 8 June 2011.

19. 'Opposition to nuclear energy grows: global poll', *BBC World Service*, 25 November 2011.
20. 'Japan left with one nuclear reactor after shutdown', *BBC News* Asia, 26 March 2011; Wieland Wagner, 'What Future Does Nuclear Power Have in Japan', *Spiegel Online*, 1 March 2012.
21. Here and passim, see Joachim Radkau (2011) 'Eine kurze Geschichte der deutschen Atomkraftbewegung', in: *Aus Politik und Zeitgeschichte*, No. 46.
22. Lucas W. Davis (2011) 'Prospects for Nuclear Power', NBER Working Paper Series 17674; Jason Koebler 'Expert: nuclear power is on its deathbed', *U.S.News.com*, 30 March 2012.
23. Davis 'Prospects for Nuclear Power'.
24. Here and in the following, see World Nuclear Industry Status Report 2010–2011, p. 53; Margaret Ryan, 'New reactors mean new life for nuclear', *AOL Energy*, 1 December 2011.
25. According to some estimates, a single nuclear power plant costs almost twice as much as a coal-fired power plant and almost four times as much as a gas fired one. 'Undeterred by Fukushima: nuclear lobby pushes ahead with new reactors', *Spiegel Online*, 8 March 2012. John Rowe, CEO of Exelon, the largest US nuclear utility, reportedly estimated that a carbon price of $100 per ton would be needed to make new nuclear energy break even. World Nuclear Industry Status Report 2010–2011, p. 32.
26. Cf. World Nuclear Industry Status Report 2010–2011, p. 29.
27. Here and in the following: John C. K. Daly, 'French nuclear anxieties soar after Fukushima', *The Huffington Post*, 16 February 2012.
28. 'French nuke industry struggles to boost public image,' *Homeland Security News Wire*, 22 September 2011.
29. Daly 'French Nuclear Anxieties'.
30. Website of the Center for Climate and Energy Solutions, http://www.c2es.org/business/belc/members/areva, accessed 20 February 2012.
31. Nathan Myhrvold 'After Fukushima: now, more than ever', *The New York Times*, 2 December 2011; World Nuclear Industry Status Report 2010–2011, p. 54.
32. Shifra Mincer, 'Is nuclear power being phased-out', *Aol Energy*, 8 December 2011.
33. World Nuclear Industry Status Report 2010–2011, p. 55, 10.
34. Leslie Hook, 'China nuclear protest picks up steam', *Financial Times*, 29 February 2012, p. 5.
35. Cf. Jim Hoagland, 'Nuclear energy after Fukushima,' *The Washington Post*, 6. October 2011; Henry Sokolski, 'Nuclear power goes rogue: Post-Fukushima, the market for nuclear power is changing latitudes', *The Daily Beast*, 28 November 2011.
36. Website of the European Nuclear Society, http://www.euronuclear.org/info/encyclopedia/p/pow-gen-ger.htm, accessed 2 March 2012.
37. Ulrike Fokken 'Back to basic economics', *The German Times for Europe*, November 2007.
38. 'Big money needed for German energy transition', *World Nuclear News*, 22 September 2011.
39. Jennifer Morgan 'In Germany's nuclear phase-out, renewable energy plans are clear', Website of the World Resources Institute, 7 June 2011, accessed 12 December 2011.

40. Gerrit Wiesmann, 'Germany faces "Herculean" task with move to renewables', *Financial Times*, 5 December 2011.
41. Sylvia Pfeiffer and Gerrit Wiesmann, 'Nuclear fears spur Berlin to bet big on renewables', *Financial Times*, 9 June 2011.
42. 'German Environment Minister: "Our lifestyle has revolved around a dangerous egotism,"' *Spiegel Online*, 28 November 2011.
43. Bundesministerium für Umwelt, Naturschutz und Reaktorsicherheit (2011) 'Renewably employed. Short and long-term impacts of the expansion of renewable energy on the German labour market.'
44. 'Aufbruch in ein neues Energiezeitalter: Gemeinsam auf dem Weg in eine nachhaltige Moderne.' Public address delivered by Environment Minister Dr. Norbert Röttgen at the Free University of Berlin on 7 June 2011 (Berlin: Bundesministerium für Umwelt, Naturschutz und Reaktorsicherheit).
45. 'A Green future for Europe's biggest economy: What Germany must learn from Chernobyl and Fukushima.' An editorial by German Environment Minister Norbert Röttgen, *Spiegel Online*, 27 March 2011.
46. 'The energy revolution in Germany can be a model for others', *The German Times for Europe*, June 2011.
47. Cf. Helmut Weidner and Lutz Mez (2008) 'German climate change policy: A success story with some flaws', *The Journal of Environment Development* 2008, 17(4), 356–378, p. 357.
48. Gabriele Hecht (1998) *The Radiance of France: Nuclear Power and National Identity after World War II* (Cambridge: MIT Press).
49. Peter Diehl, 'Uranium mining in Eastern Germany: The WISMUT legacy', http://www.wise-uranium.org/uwis.html, last updated 17 April 2011.
50. Cf. Alice Holmes Cooper (1996) *Paradoxes of Peace: German Peace Movements since 1945* (Ann Arbor: University of Michigan Press).
51. Robert Moeller (2001) 'Remembering the war in a nation of victims: West German pasts in the 1950', in: Hanna Schissler (ed.) *The Miracle Years: A Cultural History of West Germany 1949–1968* (Princeton: Princeton University Press), pp. 83–109.
52. Gudrun Pausewang (2003/1983), *Die letzten Kinder von Schewenborn: oder ... sieht so unsere Zukunft aus?* (Ravensburger Buchverlag); Pausewang (2012/1987), *Die Wolke* (Ravensburger Buchverlag). See the 2006 movie adaptation, *Die Wolke / The Cloud*, directed by Gregor Schnitzler.
53. Günther Jauch talk show, *Das Erste*, 4 March 2012, 21:45 PM.
54. Cf. Christian Joppke (1995) *Mobilizing Against Nuclear Energy: A Comparison of Germany and the United States* (Berkeley: The University of California Press), p. 109 ff.
55. Joppke, *Mobilizing Against Nuclear Energy*, p. 113. Cf. Michael Sailer, 'Nuclear waste repository case studies: Germany', *Bulletin of the Atomic Scientist*, 29 August 2008.
56. See Arne Jungjohann's chapter in this volume.
57. C. Greipl (2004), 'Public communication challenges in a federal nuclear regulatory system', in: *Nuclear Regulation; Building, Measuring and Improving Public Confidence in the Nuclear Regulator*, Workshop Proceedings, Ottawa, Canada, 18–20 May 2004.
58. Here and in the following, see: *Nuclear Legislation in OECD Countries: Regulatory and Institutional Framework for Nuclear Activities, Germany* (2011), OECD.

59. 'Gesetz über die friedliche Verwendung der Kernenergie und den Schutz gegen ihre Gefahren', website of the Federal Ministry of Justice, http://www.gesetze-im-internet.de/atg/index.html, acessed 27 March 2012.
60. Here and passim, see Radkau (2011).
61. Cf. Gerard Braunthal (1998) 'Opposition in the Kohl era: The SPD and the Left', in Clay Clemens and William E. Paterson (eds) *The Kohl Chancellorship* (Portland: Frank Cass), pp. 143–162, at p. 149.
62. Loren R. Cass (2007) 'Measuring the domestic salience of international environmental norms: climate change norms in American, German, and British climate change debates', in Mary E. Pettenger (ed.) *The Social Construction of Climate Change: Power, Knowledge, Norms, Discourse*, (Aldershot: Ashgate), pp. 36–40.
63. Weidner and Mez 'German climate change policy: A success story with some flaws', p. 357.
64. Nuclear Power in Germany; http://www.world-nuclear.org/info/inf43.html (Feburary 2012).
65. Cf. Paul Hockenos (2008), *Joschka Fischer and the Making of the Berlin Republic: An Alternative History of Postwar Germany* (Oxford: Oxford University Press).
66. Axel Vorwerk, 'The 2002 Amendment to the German Atomic Energy Act Concerning the Phase-Out of Nuclear Power,' http://www.oecd-nea.org/law/nlb/nlb-69/nlb69-vorwerk.pdf.
67. English translation at: http://www.bmu.de/english/renewable_energy/doc/3242.php.
68. Dave Graham, 'Thousands surround Merkel office in nuclear protest', *Reuters*, 18 September 2010.
69. 'Reversing the atomic phase-out: German minister sparks government row over nuclear power', *Spiegel Online*, 8 February 2010; '14 years longer online: Merkel's government extends nuclear plant lifespans', *Spiegel Online*, 6 September 2010.

7

Ambitions and Realities of the German Energy Revolution

Brian Marrs

Germany has taken center stage in the global endeavor for clean energy, rendering the country to face a litmus test for effective decarbonization policies. More than any other highly developed industrial nation, Germany has ambitiously attempted to answer three of the 21st century's defining questions. First, what should power a world confronted by climate change if energy must be clean, affordable, reliable, and publically accepted? Second, what should produce electricity if coal is too dirty, nuclear too dangerous, gas too expensive or unsecure, and renewables too costly or insufficient? In addressing the above questions, a third emerges: can Germany – and other countries – reconcile ecology with commerce while maintaining an innovative three-trillion-dollar, industrial-based economy?

Tackling the above challenges most likely requires a radical transformation of the centralized bulk power system toward a flexible, more decentralized model ultimately capable of adapting to the intermittent electricity generation from renewables, such as wind or solar power, along with other distributed low-carbon energy sources. Germany's quest for an energy revolution (*Energiewende*) is the world's largest and most audacious clean energy experiment. Moreover, as it exists today, the *Energiewende* is an unparalleled economic bet on renewables designed to address climate change. No economic power in world history has committed as strongly to renewable energy development as Germany. The country stands alone among G20 nations as the only member that aims to seriously limit or entirely phase-out both coal-fired and nuclear power while reducing greenhouse gas (GHG) emissions 40 per cent by 2020.[1] No G20 country with Germany's decarbonization targets has committed to reducing emissions without nuclear power.

Such ambition could come with high economic costs and surely pushes against the conventional financial and technological boundaries of the energy industry. Surely, many German companies look to the *Energiewende* as a golden opportunity, especially renewable energy businesses in the *Mittelstand*.[2] However, for most of Germany's largest exporters and heavy-industry, the current policy structure of the *Energiewende* – not the broader goal of decarbonization – raises alarm. The German business community frequently complains that the taxes and regulations in place from environmental policies choke economic growth and erode Germany's global competitiveness in high-quality manufacturing. The former CEO of one of Germany's largest power companies, Jürgen Grossmann, asserted (with considerable hyperbole) that Germany has become an 'eco-dictatorship', incapable of supplying affordable energy to the country's industrial base. Another prominent German corporate leader, CEO Heinrich Hiesinger of ThyssenKrupp, described the country's decarbonization policies as 'an effort comparable to reunification' in terms of financial costs and political dedication. Charged rhetoric aside, for many in Germany's business sector, the perceived price of decarbonization is deindustrialization, with the German power sector undergoing unprecedented changes.

Based on professional experience working with leadership in the German and US power industries, this author finds that many aspects of the German energy revolution toward renewables and beyond nuclear and coal power are technologically problematic, which has led to instances of suboptimal climate policies. Moreover, current energy policies contain contradictory and perhaps even disabling financial architecture. As appealing as dropping nuclear and fossil fuels might sound, any energy future beyond the conventional will require unprecedented technical advancements occurring in well-governed markets backed by long-term political resolve and financial resiliency. In the wake of the nuclear phase-out, it appears that Germany is set to replace fission with lignite, the dirtiest of coals, not renewables. In other words, while many claim that Germany's historic reliance on coal is over, the facts seem to suggest otherwise. The German Federal Network Agency (*Bundesnetzagentur*) reports less than 2 GW of coal power retiring over the near-term, while another 8 GW will commence construction, with more to come.[3] While new coal plants are more efficient than those being replaced, they will still rapidly increase Germany's GHG emissions, especially as zero-emission coal technologies, such as carbon capture and storage (CCS), remain largely untested and receive little policy support in Germany or elsewhere. At least for now, coal's position seems even more secure if Germany cannot establish natural gas and power capacity markets for incentivizing increased gas-fired generation.

More generally, the current political debate in Germany and the *Energiewende*'s compromises seem too rigid and too many: decarbonization versus deindustrialization, nuclear and coal versus renewables, and ecology versus economy. As it stands today, the *Energiewende* must evolve as a framework for broader cooperation and innovation, or risk failure. Cooperation and innovation require a careful understanding of context, which is what this chapter seeks to provide for energy business leaders and policymakers outside of Germany. Like most capital-intensive industries, change in the energy sector is slow, but should never be underestimated. Alexander Graham Bell would hardly recognize the modern telephone, but Thomas Edison would easily find today's centralized power system relatively similar to that born in 1882 at his Pearl Street Station. If the 20th-century's energy system originated in New York City, the 21st century's could easily come from Berlin, even despite some climate policy shortcomings discussed in this chapter.

This chapter seeks to explain why and how Germany's ambitious energy revolution necessitates investments of political, financial and technological capital on an unprecedented scale. To do this, this text presents the facts and figures of Germany's energy economy, examines the ambitions and related progress of German clean energy policies and finally, assesses the feasibility of Germany's energy revolution.

Energy and the German economy

Home to the world's fifth largest economy annually worth over $3 trillion, Germany produces automobiles, chemicals, machinery and power generation equipment iconic of the world's best-quality industrial goods. Only toppled by China for the title of the world's leading exporter by value, Germany accounted for roughly 9 per cent of world trade in 2008 and posted a trade surplus exceeding $200 billion. BASF, Bosch, Volkswagen, Siemens, Bayer and countless other German companies are globally omnipresent technology giants. Feeding the German economy demands energy – lots of energy. Germany is the world's fifth largest energy consumer and typically ranks as the world's fifth or sixth largest emitter of greenhouse gases. Nearly 60 per cent of Germany's total primary energy supply (TPES) is imported, meaning, Germany has few indigenous energy resources. Germany hosts Europe's largest electricity market and second largest natural gas market. This section explores the relationship between the economy and the geological, economic and political dynamics of German energy.

Outside modest deposits of hard and lignite coal, uranium and natural gas, Germany possesses few conventional energy reserves. German hard coal production is economically uncompetitive without government support against cheaper coal imports. When subsidies for hard coal end in 2018, the industry faces probable collapse.[4] Although Germany is among the world's largest lignite coal producers, harmful air, water and climate impacts associated with lignite extraction and combustion create uncertainty around future brown coal production. Especially as demand for coal increases in the wake of the nuclear phase-out, Germany has increasingly relied on imports of Polish and waterborne coal. Although Germany possesses large uranium resources, legislation ordered the cession of all uranium mining following reunification in 1991.

Natural gas is likely the optimal fuel source for cost-effectively meeting Germany's increasingly ambitious environmental targets. Containing roughly half the carbon dioxide emissions per kilowatt-hour (kWh) as coal-fired power, natural gas-fired electricity boasts inherently flexible dispatch capabilities suited for an energy system planned around intermittent generation from renewables.[5] Domestic natural gas production still provides about 12 per cent of German consumption, although conventional reserves should reach exhaustion within the next 10 years.[6]

Hydraulic fracturing, or 'fracking', combined with multi-directional drilling technologies could unlock vast natural gas production from deep shale formations. ExxonMobil, ConocoPhillips and other firms have begun exploratory drilling in Lower Saxony. Still, Germany's unconventional gas reserves remain relatively unexplored.[7] The environmental impacts of shale gas extraction have brought the viability of this resource into question across Europe, including Germany, due to public dissention and forthcoming regulations.[8] Beyond environmental concerns, Europe's restricted pipeline access, strict wildlife protection laws and mineral compensation regimes have contributed to slowing shale gas development. The shale gas revolution that has fundamentally changed energy in North America seems unlikely to replicate itself in Germany or anywhere else in continental Europe anytime soon.

With few abundant, cheap and environmentally favorable domestic energy resources, 'doing more with less' is the modus operandi of German industry. Germany is one of the world's most efficient industrial economies, earning roughly \$6,400 per ton of oil equivalent (TOE) at full market values, compared to \$5,000 in the United States. Since German heavy industry generates a higher percentage of gross domestic product (GDP) than that in the United States, the relative efficiency of Germany industry is often up to 20–50 per cent higher than the nominal US–German $1,400 gap per TOE consumed.[9] Political moves toward decoupling economic growth from energy consumption have contributed to softening the impact of high energy costs on German industry. Looking at a post-nuclear Germany, efficiency is even more critical for furthering renewables by reducing total demand and helping less energy-dense sources of supply, such as wind or solar power, meet predicted power demand.

World-class energy efficiency helps German companies cope with high energy prices, but the German economy remains exposed to price uncertainties, much of it policy-driven. Energy-intensive manufacturing industries generate 44 per cent of revenue and 32 per cent of final GDP, the highest percentages among highly developed countries.[10] Profit margins for the cars, chemicals and high-end products synonymous with quality German manufacturing are often razor-thin, meaning that sharp spikes in energy prices are immensely destructive to industrial planning and profitability. In comparison, the United States only derives approximately 30 per cent of income from energy-intensive industries. Energy-intensive industries employ some 12.4 million Germans, or approximately 24 per cent of Germany's working population.[11]

German industry bears the highest energy costs of any heavy-industry manufacturing country in the European Union, regardless of the recent efforts during the recession to reduce energy tax burdens to control rates. Alongside direct costs, the regulatory burden from navigating a patchwork of standards, taxes, emissions trading and other mechanisms for monetizing environmental value costs businesses valuable time, management resources and in many cases, consulting and/or staff training fees. The European Union's Emissions Trading Scheme (ETS) has come under intense fire from *Mittelstand* companies for complicating financial planning and disproportionately burdening small or medium-sized businesses.[12] German businesses routinely complain that 'instrument crowding', or the coexistence of multiple, distinct environmental policies in the energy sector, directly or indirectly raises the costs of energy. From this perspective, environmental policies might play an important role in German society, but can also distort competition with questionable cost-effectiveness. Moreover, the presence of multiple green energy policies can compromise the goals and integrity of each individual program, leading to reduced cost-effectiveness and environmental outcomes.[13]

The political economy of the German energy sector is fairly unique, especially when compared to that in the United States. Since the formation of the European Coal and Steel Commission and the European

Atomic Energy Community, energy has played strongly into Germany's post-war normalization and European integration. This positioned German coal, mining, steel, chemical and power companies at the heart of pan-European policy construction. Looking beyond Europe, access to natural resources drives Germany's need for stable world commodity markets and close relationships with resource-rich nations. German corporations, often acting collectively through a *Verband*, or trade association, such as the politically influential *Bundesverband der Deutschen Industrien*, work closely with German diplomatic missions to help secure access to natural resources. In particular, Germany's relationship with Russia, Europe's largest external energy provider, is arguably the strongest among EU member states and has been of considerable importance for negotiating EU energy policy regarding oil and natural gas.

German energy companies' market capitalization hardly compares to that of the US global energy giants like ExxonMobil Corporation. A broad coalition of NGOs and cleantech-oriented *Mittelstand* combined with Germany's decentralizing parliamentary structure means large German corporations lack the lobbying potency of their North American counterparts against disfavored climate and energy policies. Traditionally defined as firms with less than 500 employees, *Mittelstand* companies constitute the vast majority of German companies.[14] As one of the strongest foundations for the German middle-class and nostalgically connected to the German spirit of innovation, the *Mittelstand* has served historically as a critical business proponent for aggressive climate and clean energy policies. Large companies, particularly export-oriented manufacturers, are certainly major voices in German climate policy debates, but large corporate influences are far less deterministic of policy outcomes compared to those in the United States. Public distrust of large corporations or 'big business', especially energy providers, has helped to mobilize a robust environmental resistance movement against utility and extractive industry interests that have opposed climate policies. While most large German companies have and continue to support climate protection goals, broader industry and political consensus regarding the framework and pace of associated policies remains elusive.

Replacing the atom, managing carbon

In the aftermath of the tragedy at the Fukushima nuclear plant in Japan, Germany's energy sector found itself at a crossroads. Public opinion swung sharply against nuclear power. One of the most polarizing issues in German politics reached a final, though controversial, climax.

In May 2011, the Green Party swept the CDU out of office in Baden-Württemberg, a state under CDU control for almost 60 years.[15] This electoral loss clearly helped to spook Berlin's center-right establishment into believing that nuclear power was a losing political issue. After an intense political drama that involved over 50,000 anti-nuclear protesters rallying on the streets of Berlin, Chancellor Angela Merkel's *Schwarz-Gelb* governing coalition struck an agreement to shut down all 19 of Germany's nuclear power facilities by 2022.[16] Seven nuclear power plants were taken offline immediately after Fukushima. The gradual decommissioning of nuclear power in the coming years will remove an additional 24 GW of installed generating capacity from the electricity supply.[17]

The backbone of Germany's electricity supplies, coal-fired power, is also in transition. Coal-fired power stations generate nearly 50 per cent of Germany's electricity. Within the next decade, more than 14 GW of installed generation capacity from aging coal-fired power plants must retire from the electricity supply system.[18] With nuclear power out of the question, options to replace coal will not come easy. Natural gas is an optimal fuel for flexible, 'quick-start' generation to match intermittent renewable generation and is more easily utilized with combined heat and power (CHP) systems, which recycle waste heat from electric generation. Yet, at least for most Europeans, natural gas is an expensive fuel subject to price volatility and increasingly sold in shifting market structures. North American shale gas resources have indirectly introduced cheaper, spot-market priced liquefied natural gas products into European markets. Most European countries, including Germany, have traditionally relied on pipe-to-pipe, take-or-pay contracts, with Russian companies as the primary suppliers. With European oil-indexed prices shifting toward gas-to-gas pricing systems, Germany, like many of its neighbors, has struggled to find preferential market structures to secure affordable, long-term natural gas supplies.[19]

Moreover, without capacity markets, Germany will likely struggle to promote non-baseload, low-utilization factor power generation capable of balancing intermittent sources.

As stated in continuous press releases since the nuclear phase-out, the German government has maintained that renewables can replace nuclear power and lay the basis for gradually phasing-out fossil-fired power. This assessment is not necessarily altogether unrealistic if considered to occur by 2050 or beyond. Wind turbines, solar arrays and bioenergy generated about 12 per cent of Germany's total energy supply and 20 per cent of Germany's electricity in 2011, an increase of several orders of magnitude since 1990.[20] For the past 20 years, the implementation of pollutant trading schemes for greenhouse gases, efficiency standards, feed-in tariffs (FiTs) and eco-taxes transformed the self-declared German 'land of ideas' into the foremost laboratory for clean energy technology. Such policy-based financial systems provide higher than market rates for renewables and associated clean energy products by allocating a value for avoided pollution to each kWh generated. By orienting Germany's domestic energy market to essentially internalize the externalities of fossil fuel consumption, policy-driven instruments have built a lucrative business model around cleaner energy technologies.

Yet, while renewables are clean compared to conventional fossil-fired generation, they are costly alternatives in the current electricity market and have vastly different performance characteristics compared to conventional power sources. Households could pay over 4–6 cents per kilowatt-hour for renewables subsidies by 2012.[21] The German Wuppertal Institute for Climate, Environment, and Energy predicts that subsidies for photovoltaic solar panels could exceed $61 billion in ratepayer contributions when totaled from 2000 to 2030, yet will produce less than 2–3 per cent of Germany's electricity during this same period.[22] These costs exclude the necessity for additional transmission lines to smooth the intermittent generation profile of renewables across the electricity system. Integrating renewables into the power system remains a huge challenge for the *Energiewende*, both from technological and financial perspectives.

Constraints on transmission construction might represent the largest barrier to an electricity system centered on renewables. The German Energy Agency (dena) projects that Germany will need thousands of additional transmission and distribution lines to facilitate renewable energy expansion and meet greenhouse gas reduction targets. Large-scale renewables, such as massive offshore wind parks, are planned to reduce Europe's carbon emissions. This represents a shift in the basic philosophy surrounding renewable energy promotion. Renewables are no longer beholden to the 'small is beautiful' dogma of the 1980s and 1990s; rather, industrial-scale renewables are on the march across the European Union and United States. The organization of European Transmission System Operators (TSOs) has found that the time required to site transmission is three to five times longer than siting renewable energy generation.[23] Moreover, an EU-wide analysis concluded that the average time between the start of planning and final commissioning of a power line frequently exceeds 10 years. Although the Transmission Line Acceleration Act (*Netzausbaubeschleunigungsgesetz*) passed in the German parliament during the summer of 2011 to address regulatory and legal barriers blocking transmission planning, it is unclear how effective this legislation will be at actually promoting investments averaging $8–$16 million per kilometer.

Designing policies for mobilizing financial resources into climate-friendly technologies and infrastructure continues to represent an immense task for the German political, business and energy communities. Policy architecture must be stable enough to incentivize long-term, capital-intensive investments, yet flexible enough to adapt to ever-changing economic conditions, all the while satisfying the global and local performance idiosyncrasies inherent to energy markets.

The blueprint for ecological reindustrialization

Reconciling economy with ecology has been at the core of 21st-centry German industrial policy. For the German business community, decarbonization of the German energy system must adhere to two strategic priorities: (1) stable, affordable energy supplies and (2) domination of emerging cleantech markets. German energy policy embraces the 'precautionary principle' (*Vorsorgeprinzip*), a concept that holistically evaluates the externalities of energy production and inserts these unaccounted-for costs into the market. Constructing domestic demand for clean energy products via policy-driven mechanisms has enabled large-scale renewables deployment and 'learning-by-doing' in the innovation process. This section details how German businesses, above all the *Mittelstand*, have leveraged policy instruments to build some of the world's most impressive renewable energy companies. Renewable energy policy for solar and wind power are the primary concern of this section.

Building upon the CDU/CSU/SPD Grand Coalition's 2007 Integrated Climate and Energy Framework, the Energy Concept released by Chancellor Angela Merkel's *Schwarz-Gelb* coalition lays out the framework for Germany's climate and energy plan over the next 40 years. Not only is the German government's target of 40 per cent GHG reductions by 2020 against a 1990 baseline the most ambitious among G20 nations, it is also the only such GHG mitigation program to feature a simultaneous phase-out of nuclear power. By 2020, over 18 per cent of Germany's gross final energy consumption and 35 per cent of Germany's electricity should come from renewables. By 2050, German GHG emissions are set to fall at least 80 per cent below the 1990 baseline, while 60 per cent of gross final energy consumption and over 80 per cent of electricity supplies should come from renewable resources.

The primary facilitator of Germany's globally oriented renewable export industry rests on policy-based financial mechanisms. The

Renewable Energy Law (*Erneuerbare-Energien-Gesetz,* or EEG) guarantees renewable energy generation higher-than-market rates per kWh produced. Renewable energy generation is compensated regardless of whether each kWh actually displaces fossil generation in the electricity system. Renewable electricity also receives priority order dispatching, meaning renewables are always placed into the transmission system ahead of other power resources, regardless of their real-time marginal costs.[24] Similarly, the Renewable Energies Heat Act (*Erneuerbare Energien-Wärmegesetz,* or *EWärmeG*) increases the share of heat produced from renewables via supportive standards and market incentives. Driven by the EEG, high energy taxes, investment subsidies and generally higher consumer willingness to pay for greener energy, Germany's energy markets incubate renewable energy businesses uncompetitive in most other countries. Although it is difficult to accurately assess the climate benefits of renewables in the presence of the European Emissions Trading System (ETS), the Federal German Environmental Agency (BMU) credits renewables with avoiding the emission of 107 million tons of CO_2 in 2009.[25] While it is challenging to provide the exact costs of these reductions, it is vital to note, as have economists like Joachim Weimann, that the costs per ton of carbon reduced for many renewables is extraordinarily high, in some cases over $800 per ton.

Putting the climate benefits of renewables aside for a moment, over 212 companies in Germany are associated with the design, production or systems integration of renewable energy. These firms contract out work to hundreds of mechanical, electrical and other specialized engineering firms. Nearly 40 per cent of these renewable energy companies achieve export quotas between 30 and 40 per cent.[26] In 2008, renewable industries employed over 300,000 Germans. Roughly two-thirds of these jobs are directly attributable to the EEG. Costs of the EEG to all German consumers amount to approximately 3 per cent of household bills, about $4 per month.[27] Notably, these costs are set to rise steeply after 2012 as new solar and offshore wind installations start receiving EEG compensation. The extent to which Germans will accept paying an additional $50–$75 annually to support these renewables remains a somewhat open question for policymakers and businesses. Renewable support mechanisms in other European countries, the United States and Australia have imploded after ratepayers revolted against increased charges.

Irrespective of renewables' burden on ratepayers, business development has spurred spectacular technical progress. If the *Energiewende* is to succeed, this progress must continue. 'Demand pull' and 'learning-by-doing' have proven essential for renewable energy innovation. Technological progress has improved the economics of solar and wind power. Solar photovoltaic (PV) systems achieved incredible economies of scale with modularity, cost-reductions from mass production, and world-beating balance of system costs.[28] Standard rooftop solar PV modules are 60 per cent cheaper in Germany today than in 1990. For each doubling in installed capacity, the associated price drop for solar systems has historically been around 20 per cent in Germany.[29] By the end of 2009, global PV installed capacity reached 21 GW, of which 9.8 GW was installed in Germany. By comparison, the United States has only 2.5 GW of installed solar capacity as of 2011, despite vastly superior solar resources. Germany receives the same annual solar insolation as Alaska, whereas many heavily populated areas along the US Pacific Coast can receive 350 days of sunlight.[30]

Wind power development has followed a somewhat similar pattern of improvement as solar PV. Approximately 21,200 wind turbines with an installed capacity of 26 GW were installed in Germany by the end of 2009. Wind represents roughly 7 per cent of German electricity generation. Germany maintains 16 per cent of global installed wind power capacity, second only to the United States' 22 per cent share in global wind capacity.[31] Significant cost reductions for wind power correspond to increased turbine size and height. Electricity production efficiencies of turbines has increased 2–3 per cent annually for the past 15 years, leading to an overall cost reduction of approximately 30 per cent for turbine production capacity.[32] Over the same time period, turbines increased in size from 600kW to over 5MW. Turbine enlargement has boosted energy yield per square meter of turbine blade area 5 per cent annually over the past 15 years, an impressive engineering feat.[33]

Renewable energy is a business movement for many Germans, and renewable energy enjoys widespread popular support. A poll in 2006 found that over 65 per cent of Germans believe that Germany should increase renewable electricity generation. Roughly 80 per cent of respondents agreed the German government should increase funding for renewable energy research, development and deployment.[34] Since the liberalization of the German electricity market at the turn of the century, more than 135 companies market 'green power' renewable electricity directly to consumers. Greenpeace Energy and Lichtblick are independent ventures each attracting a fast-growing customer base; Lichtblick alone has over 100,000 consumers concentrated in Northern Germany and the Hamburg area. Approximately 560,000 Germans, roughly 1.5 per cent of the residential electricity market, purchase green power. Opinion polls consistently show that 20–35 per cent of German consumers indicate a high willingness to pay for renewable energy, though paradoxically the green power market has not expanded anywhere close to levels suggested by polling.[35] One explanation for this is that perhaps while Germans are willing to show support for renewables, few are willing to actively seek out higher electricity prices.

Whether starting an independently branded renewables company like EnBW's YelloStrom or offering premium packages of renewable-only electricity, Germany's largest electricity companies have spent over $100 million annually on green power and renewables marketing since the late 1990s. This spending is set to increase.[36] The European Union electricity providers' trade association, Euroelectric, has set a target of 80 per cent reduction of GHG emissions against a 1990 baseline.[37] In order to achieve this goal, power companies must dramatically increase green power marketing, GHG offsetting and engage consumers as to the costs and benefits of low-carbon electricity production. With the nuclear phase-out now a reality for German power companies and the push toward renewables unavoidable for large incumbent utilities, a new era of consumer outreach in Germany has begun for power companies. As of 2010, Germany's largest four utilities owned 80 per cent of fossil-fired but only 6.5 per cent of renewable generation. Interestingly, private citizens owned 40 per cent of Germany's 53GW of installed generation capacity, many through private cooperatives with buy-ins as low as $120.[38] For Germany's largest utilities, reaching green-minded consumers (and voters) could represent a difficult task given the provenance of abundant municipal- and community-based options.

No easy answers to nuclear and fossil-fired power

Mobilizing financial and political resources to support new or emerging technologies has made Germany one of the world's foremost case studies in industrial policy. Germany must devise policies to replace nearly 30+ GW of installed generation capacity phased-out or otherwise set to retire, while providing industry with cheap, reliable power and continuing Germany's ecological reindustrialization. The unavoidable necessity of replacing energy infrastructure has accelerated political pressure to decarbonize Germany's energy system. Since typical fossil-fired or nuclear power plants operate for 30–50 years, today's energy decisions set decade's long trajectories for the economy and environment. Power plants, buildings and energy-consuming or energy-producing heavy infrastructure rotate stock over the course of generations. Regulatory framework must be sufficiently longsighted to mitigate investment risks, while at the same time nimble enough to promote technology choices capable of addressing today's economic and environmental goals. Multi-billion-euro financial decisions require high levels of regulatory certainty – the kind of clarity climate policy has regularly failed to deliver, not just in Germany, but around the world.

With so much money at stake, decarbonization unsurprisingly creates considerable political friction. While few economic stakeholders in Germany disagree about the necessity or even inevitability of a clean energy future, many companies and consumers differ on the policy prescriptions enabling the energy revolution. For many in the German business community, particularly those from the technically skilled *Mittelstand*, environmentalism means securing competitive niche markets for decades of technology-driven prosperity.[39] Between 2000 and 2005, German firms filed over 10 per cent of the world's patents for clean energy technologies.[40] One of Germany's leading consultancies, Roland Berger, estimates that the global cleantech market should exceed $4.7 trillion in value by 2020, though this figure appears to assume international commitments to addressing climate change that as of this writing look rather doubtful.

Nevertheless, supporters of the *Energiewende* maintain that cleantech markets will create millions of new German jobs. Yet, regulatory instruments designed to foster cleantech contribute to Germany's rising energy costs, which already are among the world's highest, as shown in Figures 7.1 and 7.2. Industry leaders fear more expensive energy erodes German competitiveness in a global marketplace increasingly commanded by dynamic, fast-growing firms from developing economies.[41] Regulatory constraints and geopolitical shifts have driven up the comparative costs of energy in the German economy. Environmental and energy policies are not entirely to blame for Germany's rising energy costs, but it is clear that at least over the short-term, decarbonization policies will carry heavy short-term financial burdens, with longer-term benefits dependent on steady technological innovations in cleantech.

If a global *Energiewende* toward cleaner energy does occur as rapidly as many consultants and politicians claim, it is unclear whether Germany's investments in clean energy will position the economy to reap the benefits. The Chairman of the China Office at the German-Asian Business Association (*Asien-Pazifik-Ausschuss der Deutschen Wirtschaft*), Dr. Jürgen Heraeus, reflected such concerns in October 2010 when he stated in the *Financial Times Deutschland* that, 'German companies are in danger of being pushed ever higher on to the technological ladder, until one day the market niche will be too small to survive.'[42] Heraeus' view resonates

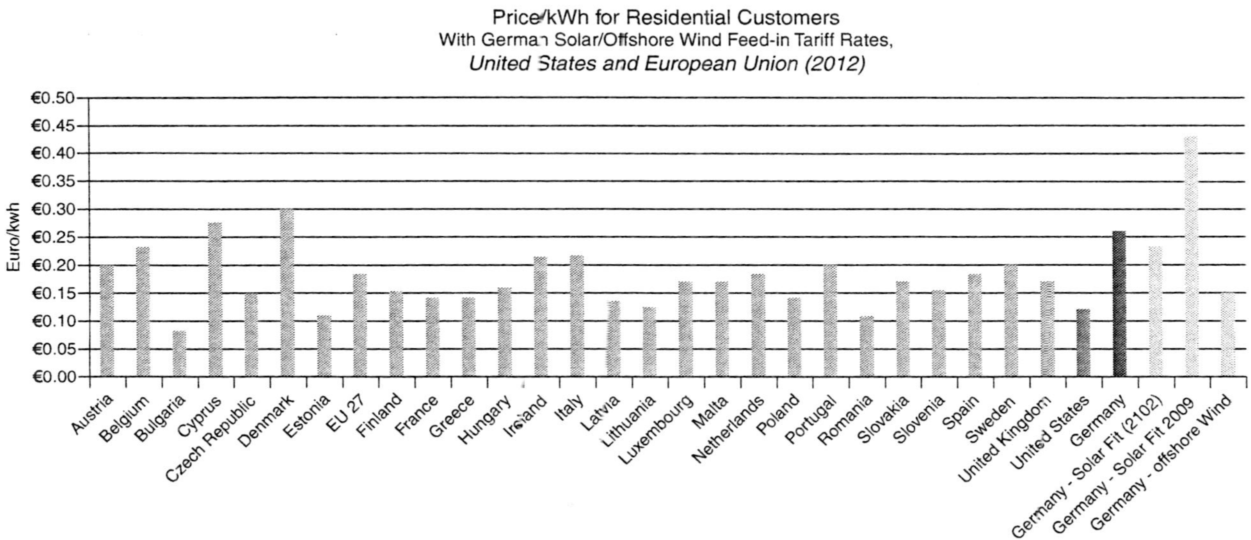

Figure 7.1 German residential power prices (2012)

Note: In some EU Member States, including Germany, taxation can represent 50% or more of the residential electricity price.

Source: *Eurostat, EIA*.

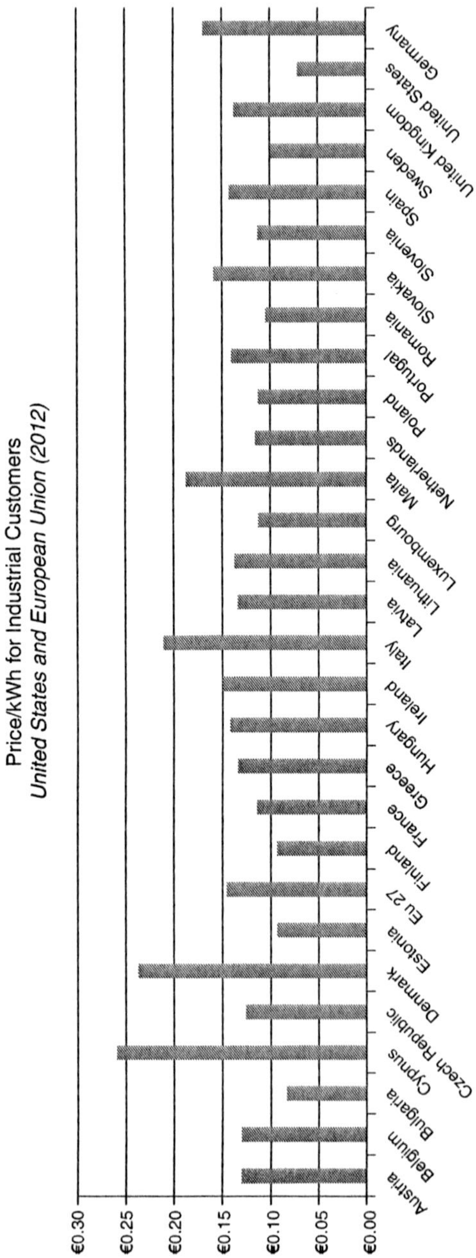

Figure 7.2 German industrial power prices (2012)
Source: Eurostat, EIA.

with an increasing number of German businessmen who worry that specialized clean energy technologies subsidized by taxpayers will be manufactured and eventually designed and/or copied by Asian competitors, particularly Chinese companies. Along with the rest of Germany's industry, German renewable companies must now compete head-on with China, a country that has managed to craft a fast-growing clean energy market virtually from scratch.[43]

Coping with competition

Low-cost labor, high-quality research facilities, promotional policies for renewables and low-carbon fuels to drive internal demand, and special 'testing' zones for clean energy technology makes China an attractive destination for low-carbon energy investment. China plans to directly invest $440 to $660 billion in clean electricity research over the next decade, a sum that is in addition to the $177 billion in energy investment included in the Chinese stimulus bill in 2008.[44] The Chinese city of Baoding has been labeled 'Electricity Valley' to purposely mimic 'Silicon Valley'. The tech cluster boasts over 200 renewable energy companies. Renewable energy manufacturing initiated under expensive public support programs in California, Germany and Spain, such as solar PV, has moved to lower-cost China.

The interim Director General of the International Renewable Energy Agency (IRENA), Hélène Pelosse, has stated that China has overtaken Germany and other major renewable energy product producers in the West.[45] The transition of solar technology from novelty to standardized production has allowed China to quickly dominant manufacturing with its low 'China price' scalability. Bloomberg New Energy Finance reported recently that, '... the Chinese in just the last two years have made the kind of progress that it took the Europeans five years to achieve'.[46]

The pace of Chinese clean energy development is staggering. In 2003, China had virtually no solar industry, but in only three years, China manufactured more solar cells than any other country.[47] China houses more than half of the world's solar cell production capacity.[48] Chinese solar PV products, manufactured by firms such as Yingli Green Energy, can produce panels over 20 per cent cheaper than German counterparts. TÜV Rheinland and like consumer protection firms approve solar panels from numerous Chinese manufacturers. Exports remain the focus for Chinese solar manufacturing growth. Most Chinese suppliers export almost 90 per cent of production, although Chinese cities like Beijing and Shanghai have developed ambitious programs to create domestic demand for Chinese solar panel production.[49] China has pledged that 15 per cent of its electricity should come from renewable sources by 2020, though such an ambitious target has not increased domestic deployment of solar PV.[50]

The superiority of the 'China price' has rocked Europe's renewable industries. Between 2008 and 2009, German solar companies saw their margins decline by around 8 per cent on average. A flood of cheap Chinese solar panels contributed significantly to the disastrous collapse of the Spanish solar market in 2009. Spain's solar market disintegration caused a panic across Europe that public support schemes could not afford a wave of renewable installations. Like that in Italy, the Netherlands, Spain and the United Kingdom, the German government moved to reduce FiT compensation for solar PV. The broad rationale behind Germany's early investments in renewables was a first-mover advantage in future cleantech markets and domestic jobs. China's renewables prowess could bring both of these aspirations into question. At least on the surface, it appeared that Germany had a first-mover disadvantage, despite billions of euros spent on publically supported schemes. If renewables fail to deliver green jobs, a critical political promise underpinning renewable policies will break down in Germany.

The limits of renewables – efficiency decides

Carbon is the heart of modern industry. Removing carbon from the industrial system is the economic equivalent of performing a heart transplant while the patient is still awake. Without delicate and adaptive policies, the pace and intensity of emission reduction targets risk sending the German economy into cardiac arrest. Unlike those in the United States, all German political parties and mainstream businesses publicly recognize that climate change is an anthropogenic problem that necessitates immediate action to mitigate CO_2 emissions. This consensus has translated into action, yet energy policy in Germany has become rigid and increasingly blind to cost-effectiveness. If the energy revolution is to reach fruition, attention toward limited financial and political resources must come into focus.

Political communication from Germany's liberal parties (Greens, SPD) often disregards the technological limitations of renewable energy and frequently implies 'fossil-fuel' interests obstruct a Germany 100 per cent powered by renewable energy.[51] For many German environmentalists, renewable energy development borders on a movement akin to a religious crusade. Although political rhetoric has shifted somewhat since Fukushima for Germany's center-right parties, conservative parties (CDU/CSU/FDP) have traditionally maintained that the costs of decarbonization could deindustrialize Germany, hurting the economy during a time of unprecedented crisis in the Eurozone. Focusing only on the costs of the *Energiewende* ignores the costs of inaction and long-term benefits received from improved air quality, climate change mitigation and the potential for future industrial growth. Though middle ground in the current political debate is unfortunately thin, it ought to occupy a position that concentrates on comprehensive energy reform. The status quo is quickly turning in to an expensive odyssey toward massive renewable projects, like offshore wind.

German electricity consumers pay an extra $10.5 billion a year in utility bills via the EEG to support renewable energy projects, which will only increase as renewables, especially solar and offshore wind power, are brought into the power system.[52] According to energy industry estimates, full auctioning of the European carbon allowances (EUAs) post-2012 will add billions of euros in extra costs passed on to German ratepayers. Accommodating intermittent renewable power sources requires building new, highly adaptive transmission systems that could cost over $70 billion in the coming decades. This timeframe is likely to be longer as public resistance has traditionally stalled transmission projects.[53] The German Energy Agency (*Deutsche Energie-Agentur*) estimates that over 3,600 km of transmission lines are required to meet a renewables target of 35 per cent by 2020.[54] During the next 12–15 years, the German government expects that industry will have to invest over $400–$500 billion into climate-friendly energy and transportation technologies.[55] Reflecting upon the enormity of this investment challenge in an interview with *Der Spiegel*, the director of the German Energy Agency, Stephan Kohler, remarked, 'The reality is that we'll need conventional power plants until at least 2050, even if we do create massive renewable energy sources. Many people dispute this. They say that we could replace power plants operated with fossil fuels by adding more renewable energy sources. My response to them is: It won't work.'[56]

High infrastructure costs translate into energy price increases. An anonymous survey of 200 energy-industry officials in Germany conducted by the Center for European Economic Research (*Zentrum für Europäische Wirtschaftsforschung*) revealed that 45 per cent of energy experts believe that wholesale power prices will rise from 4 to 6 euro cents per kilowatt-hour (kWH) during the next five years. In 2011, households could pay an extra 4 euro cents per kilowatt-hour for EEG subsidies, an increase against 2.05 euro cents in 2010.[57] Solar PV costs have become problematic. According to the Wuppertal Institute for Climate, Environment, and Energy, solar PV produces only 1.1 per cent of Germany's electricity, but receives 40 per cent of total EEG compensation. Although the German government has reduced subsidies for PV beyond 2011, the rapid increase in PV installations during 2010 will boost the cost of feed-in tariff 72 per cent.[58] Justifying PV based on climate protection is also difficult; the International Energy Agency estimates that carbon avoided via PV in the German market might cost between $500 and $1,500 per ton. By comparison, prices for authentic, certified carbon offsets in China could cost as low as $5 per ton.

The relationship between renewable energy and the performance requirements of the power system remains disjointed in public discussion, as highlighted by the nuclear debate. As of today, there is no clear path for renewables to quickly and easily displace nuclear power in Germany. A typical wind turbine generates full nameplate capacity about 20 per cent of its operational lifetime, meaning that a 3 MW turbine actually produces only a fraction of its rated capacity over its life span. Nuclear power plants avoided the emitting of 150 million tons of fossil-equivalent CO_2, an amount roughly equal to Germany's annual emissions from traffic. Over 55,000+ wind turbines would be required to even begin displacing Germany's fleet of 17 nuclear plants, especially if wind turbine capacity factors do not rapidly increase. Germany's onshore wind sites are relatively saturated, which has pushed developers to look for windier offshore sites. Germany's quest for 25 GW of installed offshore wind power will cost over $1 billion. Thus far, offshore wind faces almost debilitating technical and financial challenges. Projects are years behind schedule and running drastically over budget.[59] There are simply no easy answers for phasing out nuclear power while maintaining GHG emissions reductions from the energy sector.

A troublesome paradox has emerged in which public expectations, political rhetoric and technological realities are not aligned and counterproductive. Germany finds itself reaching for 21st century environmental targets largely with 20th-century technology. Expensive public subsidies have spurred impressive energy innovation, yet renewables alone remain technologically inadequate to meet the needs of an industrial economy. The politicization of energy in general, but especially renewable energy, has caused much of the German public to suspect that big business interests are the sole obstruction to a 100 per cent renewable future. Dr. Volker Hauff, Chairman of German Council for Sustainable Development and Advisor to the Federal Chancellery, has stated that Germany must hedge its clean energy strategy away from overreliance on renewables to achieve environmental targets. Distributed combined heat and power, demand response, efficiency, smart grid technologies, sustainable city planning and otherwise cleaner use of fossil energy, particularly natural gas, can help bridge the economic gap towards a cleaner future, with meaningful carbon reductions using today's technologies and zero-carbon with tomorrow's.

Considering the costs of low-carbon supply-side resources, the role of efficiency and demand control remains critical for the *Energiewende* to succeed. The official slogan of the Germany Energy Agency (dena) is '*Effizienz entscheidet*', or 'efficiency decides'. This slogan appropriately reflects Germany's ambitious energy efficiency standards and retrofit programs to radically reduce the country's primary energy consumption. Efficiency is perhaps the decisive, most complex, element for reconstituting a cleaner energy system, and demand-side aspects of the energy system are often the least explored, especially in North America. Despite notable successes, Germany's efficiency policies contain contradictory elements and ineffective auditing controls, as noted by an open letter to the government by 40 prominent economists published in *Die Zeit* during January 2012.[60] Even so, this author is confident that the government can address these concerns and maintain most of Germany's efficiency targets. As emphasized in the *Die Zeit* letter, and also seconded by this author, efficiency is not in itself sufficient for climate protection, but will serve as an essential compliment for Germany's ambitious supply-side cleantech targets.

In this light, the dena's slogan applies more broadly to the traditional brilliance of the German energy system. Instead of capital investments in raw energy resources and lower nominal energy prices, capital resources are expended into advanced materials and urban design to conserve resource growth. The period 1990–2006 saw an average improvement in specific energy consumption (energy intensity) of 1.7 per cent annually. German energy productivity increased 40 per cent from 1990 to 2008.[61] Impressively, German companies and research institutions submitted between 30 to 40 per cent of global patent applications in efficiency product development in 2008 alone.[62] High power prices and expensive, imported fuels have long conditioned the German industry to pursue efficiency, but price signals alone are not responsible for Germany's continued efficiency improvements.

Access to capital has proven essential. Government, financial and industry stakeholders have collaborated to finance and implement efficiency improvements across Germany's building stock. State-owned 'development banks', such as the KfW Bankengruppe have played an instrumental role in capital allocation for efficiency improvements, especially for retrofitting buildings.[63] Per government mandate, the KfW issues cheap loans, typically at half the market rate for efficiency improvements in the United States. Between 1990 and 2006, KfW issued a loan volume exceeding $40 billion to over 2.5 million homes.[64] This investment reduced GHG emissions by over 1 million tons annually, while boosting thousands of construction jobs across Germany. Efficiency consistently offers the most GHG emission reductions per dollar invested in the energy sector. Germany's ability to maintain annual efficiency improvements has underwritten success in all energy policy areas, including the development of renewable energy. Moreover, the German model for efficiency has demonstrated that this 'low-hanging' fruit is capable of growing back under the right policy and market conditions.

In addition to cheap loans, efficiency standards have reinforced Germany's progress. Germany's Energy Concept seeks to nearly halve primary energy consumption by 2050 against 2008 levels. Low-energy 'lifestyles' which embrace efficient housing and commercial buildings will play a large role in achieving this reduction. A typical US family home can consume over 400 watts per square meter. By contrast, German low-energy certified homes consume only 30–70 watts per square meter. A German passive solar house, or '*Passivhaus*', reduces energy consumption further to 15 watts per square meter for heating. Communities such as Vauban in Freiburg have pursued creating a '2,000 watt per day' community, which allow residents to consume a mere fraction of the daily energy consumed per capita in North America.[65] Over 15,000 *Passivhaus* structures currently exist, almost all of which are located in Sweden, Austria and Germany. While the bulk of the German housing stock is considerably less efficient than Passivhaus standards (such low energy homes are typically only 5–10 per cent more than the cost of standard homes in Germany), these low-energy homes focus policy on what is cost-effective and technologically feasible. Accordingly, Germany will institute one of the world's largest retrofit programs aimed at housing and commercial building stock.

Conclusion: Germany's difficult successes and path forward

This chapter has provided a brief review of German environmental policy primarily from the business and power sector perspectives. As it was intended, this chapter has provided only a brief overview of how business and policy have come together in Germany to promote the innovation and deployment of low-carbon, efficient energy technologies.

Although climate policy in Germany has been relatively effective at reducing emissions before the nuclear phase-out, it is still not clear how much the German public has benefited or will benefit from expensive, policy-driven developments in clean energy. The *Energiewende*'s success under the current energy policies looks uncertain. That said, the *Energiewende* can and should thrive as a conceptual framework for sustained compromise and continued innovation.

Such a renewed framework is sorely needed. A renewables-centric *Energiewende* jump-started as the natural conclusion of the nuclear phase-out, but, as this chapter documents, this path is rife with policies, the social complexities and potential economic disadvantages, which are just gradually coming into view. Germany's emissions will certainly rise in the near-term as it appears that largely coal, not renewables, will replace nuclear power. Juggling the nuclear phase-out, increased generation from coal-fired power, renewables deployment, decarbonization targets and the threat of deindustrialization will put enormous pressure on Germany's industry, which must now deal with strong global competition and the aftermath of the Eurozone crisis. Yet, even if big business is struggling to accept or translate the *Energiewende* into an opportunity, the German public supports an energy transition, at least for now. Like all transitions, the *Energiewende* will produce winners and losers. As it has done for much of its successful technological history, Germany is betting on the future, this time with billions and billions of Euros on the table. While some G20 governments are deciding whether to enact climate mitigation policies, Germans argue about how aggressively to tackle the problem. Even if German policies need calibration, the country still hosts the world's most audacious clean energy experiment. Change is messy, but no stranger to Germany.

Countries around the world have struggled to introduce climate policies that will cost-effectively introduce cleaner energy technologies to reduce GHG emissions. Germany's experience with climate policy shows that no perfect energy technology exists to accomplish this goal, and thus political solutions to realize a cleaner, more affordable and more reliable energy system will likely be equally imperfect. The policy-learning process – like most realities in the energy sector – can last for decades. Perhaps most revealing about German energy policy is the elusiveness of a global cleantech revolution. Germany possesses education and health care systems consistently ranked among the world's best. High-quality exports and tireless technological innovation form the basis of the German economy. In comparison to nearly all democratic countries, German voters have consistently prioritized climate change and energy issues on the political agenda. If Germany cannot reasonably retool its economy to produce clean, reliable and affordable energy, there are likely few other countries in the world with the political, technological and economic resources to do so either, especially in the absence of strong global efforts to reduce GHG emissions.

Looking forward, the eyes of energy policymakers around the world are on Germany. The birthplace of much of atomic science, Germany has walked away from perhaps the most influential energy innovation of the past century – nuclear power. The revolution Germany hopes to ignite across the world is bold, unprecedented and seeks to restructure the electric power industry from one of largely centralized to decentralized production. Despite our perceptions of modern technology, most of the power system remains similar to that of the past 100+ years, namely, extracting raw materials from the earth and burning them to produce steam. The German Energy Revolution seeks to break this historic legacy. In so doing, the *Energiewende* will push the limits of finance, energy security and economic growth, but is nevertheless solidly in line with the broader constraints facing the 21st century. The *Energiewende* represents a gamble for Germany, but those choosing to ignore climate change have also chosen to gamble.

Notes

1. Federal Ministry for the Environment, Nature Conservation and Nuclear Safety (2009) *New Thinking, New Energy. Energy Policy Roadmap 2020*, http://www.germany.info, accessed 10 January 2011.
2. The German term *Mittelstand* is a difficult term to translate into English. It refers generally to small or medium-size enterprises with less than 500 employees. *Mittelstand* companies constitute the majority of German companies.
3. German Federal Network Agency, http://www.bundesnetzagentur.de, accessed 10 December 2012.
4. Craig Whitlock, 'German Hard-Coal Production to Cease by 2018', *Washington Post*, 30 July 2007.
5. Stephan Brown, Alan Krupnick and Margaret Walls (2009) *Natural Gas: A Bridge to a Low-Carbon Future* (Resources for the Future), http://www.rff.org, accessed 10 January 2011. See also: Dietmar Schüwer and Karin Arnold (2010) *Erdgas: die Brücke ins regenerative Zeitalter : Bewertung des Energieträgers Erdgas und seiner Importabhängigkeit; Hintergrundbericht im Auftrag der Greenpeace Deutschland e.V.; Endbericht* (Wuppertal Instiut für Klima, Umwelt, Energie), http://www.wupperinst.org, accessed 10 January 2011.
6. International Energy Agency (2003) *Energy Policies of IEA Countries – Germany 2002 Review* (OECD/IEA Publication). See also Christian Wüst 'Fear of fracking: Germany balks on natural gas bonanza', *Der Spiegel*, 5 October 2012.
7. 'Bubbling under', *The Economist*, 3 December 2009.

8. Frank Dohman, Alexander Jung and Jan Puhl, 'New drilling technologies shake up global market', *Der Spiegel*, 3 March 2011. See also Michael Grossman, Das groβe Rennen um Schiefergas', *Financial Times Deutschland*, 6 March 2011.
9. Energy data was retrieved from the International Energy Agency's statistics database, http://www.iea.org/country/index.asp, accessed 10 March 2011.
10. McKinsey & Company Inc. on behalf of the 'BDI Initiativ – 'Business for Climate' (2009) *Costs and Potentials of Greenhouse Gas Abatement in Germany* (McKinsey & Company, Inc.).
11. McKinsey & Company (2010) *Energy Efficiency: A Compelling Global Resource* (McKinsey & Company).
12. Alexander Jung, 'The pitfalls of Europe's new emissions trading system', *Der Spiegel*, 10 December 2010.
13. Dieter Helm (2012) *The Carbon Crunch* (New Haven, CT: Yale University Press).
14. Eric Solsten (1995) *Germany: A Country Study* (Washington: GPO for the Library of Congress), http://countrystudies.us/germany/, accessed 10 January 2011.
15. 'A green revolution', *The Economist*, 31 March 2011, http://www.economist.com, accessed 31 March 2011.
16. Each of Germany's political parties is identifiable by a color. For example, the Free Democrats (FDP), are 'yellow' or '*gelb*' in German, the Christian Democrats (CDU) are 'black' or '*schwarz*' in German. '*Schwarz-Gelb*' thus refers to a ruling parliamentary majority, or coalition, of CDU-FDP.
17. 'Merkel's government extends nuclear plant lifespans', *Der Spiegel Online*, 6 September 2010. See also 'Berlin struggles to realize nuclear-free ambitions', *Der Spiegel Online*, 12 March 2012.
18. 'Merkel's government extends nuclear plant lifespans', *Der Spiegel Online*, 6 September 2010. See also 'A revolution for renewables: Germany approves end to the nuclear era', 30 June 2011.
19. Anthony Melling (2010) *Natural Gas Pricing and Its Future: Europe as the Battleground*. Report; Carnegie Endowment for International Peace, www.CarnegieEndowment.org/pubs.
20. Germany Working Group for Renewable Energy Statistics, http://www.erneuerbare-energien.de, accessed 28 December 2012.
21. Nicholas Comfort, 'Germany's Renewable Energy Costs Threaten Public Support, ZEW Center Says', *Blomberg.com*, accessed 20 December 2012.
22. Federal Ministry for the Environment (2009) *Electricity from Renewable Energy Sources: What Does It Cost Us?* (Berlin: Federal Ministry for the Environment, Nature Conservation and Nuclear Safety).
23. European Network of Transmission System Operators for Electricity (2010) 'ENTSO-E Position Paper on Permitting Procedures for Electricity Transmission Infrastructure' (Brussels: ENTSO-E), www.entsoe.eu, accessed 30 November 2012.
24. A more detailed explanation of the marginal cost pricing and market dispatch is available from the Fraunhofer Institute. See Frank Sensfuß, Mario Ragwitz, and Massimo Genoese (2007) *The Merit-order effect: A detailed analysis of the price effect of renewable electricity generation on spot market prices in Germany* (Fraunhofer Institute for Systems and Innovation Research, Working Papers 'Sustainability and Innovation' S7/2007).

25. For more information, see: Federal Ministry for the Environment (2008) *Electricity from Renewable Energy Sources: What Does it Cost Us?* (Berlin: BMU).
26. German Institute for Economics, German Center for Air and Space Research, German Center for Solar and Hydrogen Energy, German Institute for Future Energy Systems (2008) *Economic Analysis and Evaluation of the Effects of the Renewable Energy Act*; study on behalf of the German Environmental Ministry (Berlin, Stuttgart, Saarbrücken).
27. *Electricity from Renewable Energy Sources* (2008).
28. German Federal Association for Solar Power (2011) 'Facts and Figures about the International PV Market', http://www.solarwirtschaft.de/en/photovoltaic-market.html, accessed 10 January 2011. See also: Joachim Seel, Galen Barbose, and Ryan Wiser (2012) *Why Are Residential PV Prices in Germany So Much Lower Than in the United States? A Scoping Analysis* (NREL), http://rael.berkeley.edu, accessed 6 December 2012.
29. 'Facts and Figures about the International PV Market' (2008).
30. US Solar Industries Association online portal, http://www.seia.org/cs/research/, accessed 24 May 2011.
31. World Wind Energy Association (2010) *World Wind Energy Report 2009* (Bonn:WWEA).
32. European Wind Energy Association (2011) 'Facts about Wind Energy: Trends Influencing the Costs of Wind Power', http://www.wind-energy-the-facts.org/en, accessed 10 January 2011.
33. C. Jakobsen, H. Reymann-Carlsen, J. Boogaard, A. Martin Martin, N. Kragelund and B. Balschmidt (1999) *Insurance of Wind Turbines*, http://gramjuhl.dk, accessed 16 December 2012.
34. Udo Kuckartz, Anke Rheingans-Heintze and Stefan Rädiker (2007) *Klimawandel im Bewusstsein* (Study published by the Phillips-University Marburg).
35. Rolf Wüstenhagen and Michael Bilharz (2004) 'Green Energy Market Development in Germany: Effective Public Policy and Emerging Customer Demand', IWOe Discussion Paper No. 111.
36. Wüstenhagen and Bilharz (2004).
37. Eurelectric website, http://www.eurelectric.org/CEO/CEODeclaration.asp.
38. David Buchan (2012) *The Energiewende – Germany's Gamble* (Oxford Institute for Energy Studies).
39. Bernd Venohr and Klaus Meyer (2007) *The German Miracle Keeps Running: How Germany's Hidden Champions Stay Ahead in the Global Economy* (Working Paper No. 30 of the Institute of Management Berlin at the Berlin School of Economics).
40. William Pentland, 'The top cleantech countries', *Forbes.com* 12 June 2010. See also: United Nations Environment Programme, European Patent Office, and International Center for Trade and Sustainable Development (2010) *Patents and clean energy: bridging the gap between evidence and policy* (Final Report).
41. Stanley Reed, 'Higher energy costs plaguing Europe', *New York Times*, 26 December 2012.
42. Daniel Schäfer, 'Germany's Heraeus warns on China', *Financial Times*, 5 October 2010.
43. Erik Nebel, 'World solar power leader Germany stand at crossroads', *The China Post*, 27 December 2010.

44. The Pew Charitable Trusts (2010) *Who's Winning the Clean Energy Race?* A joint publication with The Clean Energy Economy Foundation (Washington: The Pew Charitable Trust).
45. May Hanne, 'China is overtaking Germany', Interview with Hélène Pelosse, *New Energy*, http://www.newenergy.info/index.php?id=2071, accessed 10 January 2011.
46. Louis Uchitelle, 'Green economy is not yet made in U.S.A.', *The New York Times*, 22 April 2010.
47. Evan Osnos, 'Green giant', *The New Yorker*, 21 December 2009.
48. 'The rise of big solar: growing pains', *The Economist*, 15 April 2010.
49. Valerie J. Karplus (2007) *Innovation in China's Energy Sector* (Program on Energy and Sustainable Development at Stanford University, Working Paper No. 61).
50. Rob Atkinson, Michael Shellenberger, Ted Nordhaus, Devon Swezey, Teryn Norris, Jesse Jenkins, Leigh Ewbank, Johanna Peace and Yael Borofsky (2009) *Rising Tigers, Sleeping Giant* (The Breakthrough Institute and The Information Technology & Innovation Foundation).
51. '100% Erneuerbare sind das Ziel', Website of the German Green Party, http://www.gruene-partei.de/, accessed 22 April 2011.
52. *Electricity from Renewable Energy Sources* (BMU, 2008).
53. Michael Fröhlingsdorf, 'Electrical resistance: NIMBY protests threaten Germany's energy revolution', *Der Spiegel Online*, 18 April 2011.
54. Michael Fröhlingsdorf, 'Pitfalls of green energy revolution: Public resistance grows to new monster: power masts', *Der Spiegel Online*, 5 January 2011.
55. Frank Umbach (2008) *German Vulnerabilities of its Energy Security* (Essay for the joint AICGS/Institute for German Studies at the University of Birmingham conference).
56. Interview with Stephan Kohler, *Der Spiegel Online*, 16 November 2012.
57. Nicholas Comfort, 'Germany's renewable energy costs threaten public support, ZEW Center says', *Bloomberg.com*, 20 December 2010.
58. Rob Dieterich and Vince Bielski, 'Merkel's nuclear plan earns derision as clean power costs climb', *Boomberg.com*, 4 January 2011. See also: Klaus Stratmann, 'Sonnenenergie wird unbezahlbar', *Handelsblatt*, 21 June 2011.
59. Matthias Schultz, 'Germany's offshore fiasco: North Sea wind offensive plagued by problems', *Der Spiegel Online*, 4 September 2012.
60. 'Brief and Kanzlerin: Forscher warnen vor Scheitern der Energiewende', *Die Zeit Online*, 18 January 2012. See also: Malcolm Keay (2011) *Energy Efficiency – Should We Take It Seriously?* (December 2011), Oxford Institute for Energy Studies.
61. McKinsey & Company (2010) *Energy Efficiency: A Compelling Global Resource* (McKinsey & Company).
62. German Federal Ministry of Economics and Technology (2008) *Energy Efficiency – Made in Germany* (Berlin: Federal Ministry of Economics and Technology).
63. Tatijana Bruns (2008) *KfW's Best Practice Experience in Germany and its Financing and Advisory Initiative in New Member States* (Presentation at the Second European Sustainable Energy Week, 2008, Brussels, Belgium).
64. Tatijana Bruns *KfW's Best Practice Experience in Germany and its Financing and Advisory Initiative in New Member States*.
65. More information about Passivhaus standards and building design may be found at the Passivhaus Institut, Darmstadt, http://www.passivhaustagung.de/texte/Energieeffizienz.pdf, accessed 23 April 2011.

Part III
Transatlantic Cooperation at the Macro and Micro Levels

8
The Transatlantic Climate Bridge: Challenges and Opportunities

Anja Kueppers-McKinnon, Georg Maue and Carmen Krista

The concept of a 'Transatlantic Climate Bridge' spanning the geographical divide between Germany, the United States and Canada was born out of the German government's desire to add a new layer to the transatlantic dialogue. The aim of this German climate initiative, launched in 2008, was to foster on-going transatlantic cooperation and partnerships in the climate and energy arena on all levels. The idea was born at a time of growing concern in Europe about rapidly rising greenhouse gas emissions and the desire to cooperate with the United States as a partner in the fight against climate change.

International environmental policy has been a priority for the German government for many years. For this reason, Germany has set itself ambitious climate goals as part of its integrated energy and climate policy, with the aim of putting the country on a path of less energy dependence and creating a low carbon society. Germany also knows that progressive climate policies make economic sense. It has proven that reducing greenhouse gas emissions and increasing energy efficiency lead to technological innovation and increased employment. Over 380,000 people now work in Germany's renewable energy sector, a number which has grown consistently. The renewable energy sector is also attracting substantial investment – at around €30 billion in 2011, investments in renewable energy installations were around one-third higher than in 2009 (€20.7 billion).

Regardless of its own ambitious emission reductions and green job creation goals, the German government's stance is that combating global climate change can only be solved if nations work together. Given the successful cooperation between America and Europe over past decades, Berlin is confident that the common challenge of climate change and energy security can be tackled *together*, by improving cooperation through the 'Transatlantic Climate Bridge' initiative.

When the idea of building a 'bridge' to foster transatlantic climate cooperation was first suggested in 2008, however, climate and energy issues were not the top priority in Washington. The United States had not signed the Kyoto Protocol and there was insufficient interest on the federal level in developing the renewable energy and energy efficiency sectors in order to make a significant reduction in greenhouse gas emissions. Yet on a regional level in the United States, an interest in and understanding of the benefits of renewable energy and energy efficiency was spreading. Regions such as the Midwest, buckling under the economic crisis and massive job losses, saw a solution in the creation of 'green jobs' and green industries such as solar and wind energy and electric car batteries. This gave the 'Transatlantic Climate Bridge' initiative its strong regional focus. It was designed specifically to connect all those who sought to make a difference in the climate and energy arena throughout the United States and German *Länder*, and to support partnerships that could help pave the way for joint solutions on both sides of the Atlantic.

The 'Transatlantic Climate Bridge': Launched in 2008

The 'Transatlantic Climate Bridge' was officially launched in September 2008 by the German Foreign Office and the German Environment Ministry at an international conference in Berlin. Its aim was, and is, to highlight common solutions, exchange experiences on clean technology, business practices and common market conditions between Germany and the United States.

Common solutions: Increasing the amount of practical transatlantic exchange in climate change and energy security will lead to technological and policy innovations.
Common measures and targets: Working toward common standards in fields such as energy efficiency and biofuels will develop a transatlantic marketplace of ideas and sustainable products.
Common market conditions: Working together to design a common framework will facilitate transatlantic and global emissions trading.

The initiative's kick-off in Berlin was followed by a US launch in December 2008 at the German Embassy in Washington D.C. Speaking at the event, Dr. Klaus Scharioth, German ambassador to the United States from 2006 to 2011, said:

> ...by working together, Germans and Americans can be a powerful engine for transatlantic and broader climate cooperation. We both have

strong industries, and we are leaders in technology and science. We can learn a lot from each other, and together we can make a difference.

Ambassador Scharioth argued that the answer to both the climate and the financial crises is best given by creating a 'global green economy', which can generate economic growth, employment, energy security and prevent the climate from deteriorating further. Germany's own efforts to turn its economy green gives it an important story to offer other nations, and it's these examples of best practice that the 'Transatlantic Climate Bridge' aims to share through increased transatlantic communication and cooperation.

A progressive climate policy is vital for a healthy economy: The German experience

Job creation through the development of renewable energy and energy efficiency is of interest to any country, and Germany offers an example worth considering. In 2012, renewable energy boasted a share of Germany's electricity consumption of over 25 per cent. With current scenarios showing that in just 10 years renewables can cover 40 per cent of Germany's electricity supply, Germany has managed to successfully decouple its greenhouse gas emissions from its economic growth.

Renewables have increased their contribution to climate protection. In 2011, around 130 million tons of greenhouse gases were avoided through the use of renewable energies (2009: 111 million tons). Around 70 million tons of these savings can be attributed to the Renewable Energy Sources Act (EEG) alone (see below for more information about the EEG). Germany is also set to clearly exceed its Kyoto target of 21 per cent. By the end of 2011, national greenhouse gas emissions had already been reduced by around 25 per cent compared with 1990.

As part of an integrated energy and climate program and further development of the energy concept of 2010, Germany has committed to aggressive emissions reductions targets exceeding those of the European Union's:

- a 40 per cent greenhouse gas emissions reduction by 2020 and 80–95 per cent reduction by 2050
- at least 35 per cent renewables in electricity by 2020 and 80 per cent by 2050
- renewables will cover at least 60 per cent of total energy consumption by 2050

- a minimum of 14 per cent renewables in heating by 2020
- cutting the national total energy consumption 50 per cent below 2008 levels by 2050
- cutting the national electricity consumption 10 per cent below 2008 levels by 2020, and 25 per cent by 2050
- a doubling of cogeneration (combined heat and power [CHP]) up to 25 per cent of electricity generation.

The Renewable Energy Sources Act: A German success story

The German government's long-term goal is to turn Germany into one of the most efficient and greenest economies in the world by replacing conventional energy sources with renewable energies, while at the same time ensuring competitive prices and a high and broad level of prosperity. The Renewable Energy Sources Act (*Erneuerbare-Energien-Gesetz / EEG*) has played a major role in the growth of renewable energies, providing three basic elements.

First, electricity produced from renewable energy sources is given priority for grid connection. Second, grid operators are obliged to feed in electricity produced from renewable energy. Third, there is a feed in tariff (FiT) guarantee for 20 years, providing security for investors. With this, the long-term FiT provided by the EEG is the most effective measure for expanding the use of renewable energies in a cost-efficient manner. Feed-in tariff policies have been enacted in more than 60 nations around the world, including Australia, Belgium, Canada, the Republic of Korea, South Africa and Thailand, and in some states in the United States. The UK recently switched from a quota system to FiT. Nevertheless, the design of the EEG has to be adjusted to changing market conditions on a regular basis with the aim of phasing-out financial support and achieving market-readiness for renewables in the long run.

Since 1 January 2009, the Renewable Energies Heat Act (EEWärmeG) has also been promoting the increased use of heat from renewable energy sources. The economic benefits of this policy are numerous. The share of electricity from renewable energy has increased from 6.3 per cent in 2000 to over 25 per cent in the first half of 2012. In the same time, investments totaling over €30 billion have been made in Germany's renewable energies sector, and now employs approximately 380,000 people in Germany, especially in small and medium-sized companies. Brandenburg in eastern Germany is now a leading European cleantech center. Exports of renewable energy technologies have increased as well, and Germany's share in environmental technology is now more than 16 per cent of the world market.

Society also benefits from savings from reduced energy imports of traditional energy resources (oil, coal, gas, uranium) worth more than €11 billion today. These are expected to reach approximately €20 billion by 2020. Furthermore, taking into account that the expanding use of renewables also reduces external costs (which are the public costs for environmental and climate damages not included in the market) by €9 billion, renewable energies are already competitive today.

The 'Transatlantic Climate Bridge' in action: Exchanging know-how between Germany and the United States

The launch of the 'Transatlantic Climate Bridge' at the end of 2008 came at a time of optimism that the United States was ready to join the international community in committing to emissions reductions targets. Many examples of growing cooperation in the climate and energy arena were and are flourishing between Germany and the United States, especially on a regional level. Strong partnerships exist between states such as Northern Virginia and Stuttgart, and Pennsylvania and North-Rhine Westphalia. First Solar, an American solar module manufacturer with the largest 'thin film' manufacturing site in Germany, was proving that US companies could make lucrative profits by taking advantage of the stable framework guaranteed by Germany's feed-in tariff system. In 2008, the German Fraunhofer Institute for Solar Energy Systems and MIT set up the MIT Fraunhofer Center for Sustainable Energy Systems, in Cambridge, Massachusetts, and the National Renewable Energy Laboratory (NREL) in Boulder set up a cooperation agreement with the German Aerospace Center (DLR).

Germany knew how important America's leadership would be to achieve agreement at the much anticipated UN climate conference in Copenhagen in 2009, so the 'Transatlantic Climate Bridge' initiative set out to encourage US involvement by building on these existing German–US ties. The aim was, and is, to broaden the dialogue and bring together decision-makers from states and cities in Germany and the United States, from American and German companies, from the media, from universities, from research institutes and think tanks. The challenges involved are numerous. It is a challenge to argue against the deep-seated skepticism that exists amongst many in the United States regarding climate change and the need to take action to reduce greenhouse gas emissions at a time of economic uncertainty.

Germany has a different system of government compared to the United States, and the assumption made by many is that German initiatives to encourage the development of renewable energy or energy efficiency could never be implemented in the United States. Yet despite the challenges, the 'Transatlantic Climate Bridge' has engaged both republicans and democrats, both climate skeptics and those actively involved in sustainable development, in the hope of creating a constructive dialogue. By highlighting the economic benefits and job creation possibilities inherent in progressive climate policies, the idea is to offer a different view of 'green growth'.

State level cooperation

Virginia

Virginia, like Germany, still relies heavily on coal for its energy needs. Yet there is great potential for developing wind energy production off the coast of Virginia as well as an increasing interest in sustainable planning, energy efficiency and transportation management in Northern Virginia in particular. The Northern Virginia Regional Commission (NVRC) has independently built up a strong partnership with the Verband Region Stuttgart in Germany over the past decade, giving many local officials the opportunity to travel to Germany and see how the German government's climate and energy policies are being implemented.

The commitment by Germany and the Commonwealth of Virginia to cooperate in the climate and energy field was made official in April 2009 when Virginia's then Governor Tim Kaine and Germany's former federal Environment Minister Sigmar Gabriel signed a joint declaration to support the exchange and application of mutually beneficial sustainable energy and climate change policies. This declaration has been brought to life since then through a series of joint events and strengthened cooperation between the Northern Virginia Regional Commission (NVRC) and the German Embassy, as part of the Transatlantic Climate Bridge.

In March 2009, the German Embassy partnered with the NVRC to host a workshop entitled 'Developing and Implementing a Community Energy Plan for Northern Virginia'. Energy specialists Peter Garforth and Stefan Blüm spoke to an audience of leading representatives from Virginia's local, state and regional governments about developing competitive approaches to reducing the economic and environmental impact of energy use. Peter Garforth had been closely involved in the design of North America's first comprehensive community energy plan in Guelph, Canada, which has set the city on the path of using less energy in 25 years than it does today, despite expected residential growth of 65,000 people. As head of the clean energy department at MVV decon GmbH, an energy consulting service in Mannheim, Germany, Stefan Blüm presented Mannheim's example of energy-efficient city planning and was able to illustrate how this could be applied to any city in the United States.

As a direct result of this initial workshop, Loudon County in Virginia became the first county in the USA to adopt a community energy plan, with very specific measures to reduce its greenhouse gas emissions. Arlington in Virginia is now also in the process of implementing a community energy plan, and interest has been expressed by other counties in Virginia and Maryland and from the District of Columbia.

Georgia

In 2011, the Transatlantic Climate Bridge staged an exhibition on German renewable energies at the Georgia General Assembly. This happened at the same time as a joint committee hearing of the Economics and Energy Committees of the Georgia House and Senate, which was showcasing German renewable companies that had created hundreds of new jobs in Georgia.

Policy workshops with US universities

The German embassy partnered with a variety of US universities including Johns Hopkins University, American University and Virginia Tech to host a series of policy workshops aimed at presenting a transatlantic perspective on climate change issues and energy policy. Partnering with universities meant the Transatlantic Climate Bridge message would reach students and the academic community, an important audience for strengthening transatlantic dialogue and encouraging partnerships and exchange between academic and research institutions in the United States and Germany.

Partnering with universities also gave the Transatlantic Climate Bridge events access to the political community in Washington D.C. By hosting policy workshops at Johns Hopkins University's facilities in the center of town, a strong representation from Capitol Hill came to the workshops including lobbyists, staffers and politically engaged NGOs – all important multipliers for the message of transatlantic climate and energy cooperation.

Each event featured a speaker from Germany and several speakers from the United States, with topics including international climate negotiations ('Looking Ahead to Copenhagen and Beyond: Finding Common

Ground and Adjusting Expectations'), renewable energy development ('Climate Change Policy Lessons from Europe: Innovative Approaches to Renewable Energy Promotion and the EU Carbon Cap and Trade Program') and regional energy planning ('Energy Independence: A State and Local Perspective').

The speakers brought to the United States by the Transatlantic Climate Bridge for these events included experts working in the German or international climate and energy arena, who also boasted experience in transatlantic cooperation:

- Sascha Muller-Kraenner, European representative for The Nature Conservancy in Berlin, spoke about expectations for the international climate negotiations in Copenhagen in December 2009. Mr. Muller-Kraenner formerly served as the director for Europe and North America at the Heinrich Böll Foundation's (the German Green Party's political foundation) and is a member of the Working Group on Global Issues of the German Council of Foreign Relations.
- Dr. Martin Hoppe-Kilpper, formerly with the German Energy Agency (dena) in Berlin, is currently managing director of a technology-based, competence network on distributed generation (deENet) in Kassel, Germany, and spoke about the incentives for communities to engage in their own local energy planning, and reported on best practice examples from Germany.

In 2010, students from 14 universities including American University, the University of Utah, Louisiana State University and Vanderbilt University participated in the 'Green Shot' and 'Change in Your Neighborhood' campaigns, and in the 'Green Energy Alternatives Debates'. The final four winners of the contests, which were designed to encourage creativity and innovative ideas to promote climate protection, won a trip to Berlin.

Climate- and energy-focused informational visits

Since the launch of the Transatlantic Climate Bridge in 2008, dozens of American decision-makers in the climate and energy arena have traveled to Germany to experience firsthand how the German government's climate and energy policies are being implemented and the results they are delivering. Giving a group of Americans, both 'believers' and 'skeptics', the opportunity for face-to-face conversations with German policy makers and industry representatives has created the opportunity for a rare level of open communication.

Seeing firsthand the implementation of Germany's climate and energy policy, both on a federal and a regional level, offers a deeper understanding of both the challenges and the benefits, and often breaks down assumptions of what is and is not possible. The visits undertaken focused on a variety of topics within the climate and energy field including green jobs, green growth, international cooperation on climate and energy issues, sustainable architecture and a transatlantic farmers' dialogue. Each visit typically consisted of five days in two or three German cities, starting in Berlin with a variety of ministerial visits (the Federal Foreign Office, the Environment Ministry, the Ministry of Economics) to discuss the development and implementation of climate and energy policies. Participants were also invited to join transatlantic policy discussions hosted by the US embassy or at German think-tanks.

Experiencing everyday life in Germany has proven to be an important part of these trips, so visits to the Reichstag, the seat of Germany's parliament, and a 'passive house' apartment building in a Berlin suburb, offered examples of energy-efficient buildings powered by renewable energy. Participants visited renewable energy and carbon capture and sequestration (CCS) installations, illustrating how Germany is adapting and transforming its traditional industries.

Traveling through Brandenburg, a region in the former East Germany, which suffered from heavy pollution and unemployment just 20 years ago, presents the progress made in Germany's renewable energy sector: today, the area is known as Germany's 'solar valley' due to the blossoming of photovoltaic (PV) production and wind energy there. Although the global market for PV panel production suffers from over-production, which leads to the insolvency of some PV panel production plants in Germany, the majority of jobs in this sector is related to production and installation, and persist.

Other German cities and regions, such as the Ruhr, show the development of green energy and sustainable practices taking place throughout the country. The Ruhr is an area undergoing a radical transformation – once the center of Germany's coal production, it is now a hub of renewable and smart energy and serves as an excellent example for regions in the United States that are facing similar challenges.

Participants traveling to the Ruhr region met the Mayor of Gelsenkirchen in the Science Park Gelsenkirchen – a research center which is a catalyst for the structural and industrial changes occurring in the Ruhr region and which boasts Germany's first large-scale, solar rooftop installation. Participants were also offered the chance to meet the head of the 'InnovationCity Ruhr' task force, an association of

70 large international and German companies, illustrating the participating member businesses' innovation in CO_2 reduction, energy efficiency and climate protection.

The informational visits to Germany are designed to engage diverse stakeholders from a variety of US regions and from different political backgrounds. The German embassy in Washington worked with the Midwestern Governors' Association, the Southern Governors' Association and with German Consulate Generals throughout the United States to reach state and local policymakers with varying interests and priorities, but with a common interest in learning how issues such as renewable energy development, energy efficiency, emissions trading and low-carbon growth could be implemented in the United States.

The most recent trip organized by the Transatlantic Climate Bridge in November 2012 brought republican state legislators, and members of republican think tanks and organizations to Berlin and Hamburg, facilitating discussions with members of different ministries, business people, scientists and representatives of non-profit organizations.

Whilst often generating heated discussions, these visits have resulted in consistently positive feedback from the US delegations. Many suggested that the experience of traveling through Germany exposed them to completely new information and ideas and brought about a shift in perspective of what is possible:

Skip Pruss, Director and Chief Energy Officer in the Department of Energy, Labor and Economic Growth, Michigan (June 2009):

> The fact that Germany is successfully "re-industrializing" its economy holds hope and promise for the Midwestern "rustbelt" states. You provided a tremendously valuable 'emersion' in the German model for which I will always be grateful.

Andrea McGimsey, Chair of the Loudoun County Board of Supervisors' Energy Efficiency Committee, Virginia (June 2009):

> The trip to Germany ... will prove to be a great step forward in the partnership between the US and Germany on the critical issues of climate and energy. We in Loudoun County are appreciating the fruits of the partnership already, as we prepare our Community Energy Plan ... Loudoun County is one of the fastest growing counties in the nation, and if we can incorporate systems like the ones we saw in Berlin, that would be great.

Hugo V. Hodge Jr, Executive Director (CEO), Virgin Islands Water and Power Authority (June 2010):

> The information on how Germany is making their transformation was completely new. I now understand that to affect a change of this nature, it takes the entire community 'buy in'. The German citizens have a desire to bring renewables on the grid and have committed to do so even if means increased cost. We will need that same commitment in the US and I think that will be the biggest challenge.

Brad Williams, Deputy Secretary of Energy, State of Oklahoma (June 2010):

> Much of the detail of the information provided was new, and yes, it brought about a shift in perspectives. A great many ideas were generated on this trip ... the trip was easily one of the most valuable experiences that this Office has undertaken.

Mary Gade, President of Gade Environmental Group and former Director of the Illinois Environmental Protection Agency (Nov. 2012):

> The Transatlantic Climate Bridge Initiative is an outstanding program for understanding Germany's distinctive national energy policy. The information I gained will be invaluable in my environmental practice and in speaking to others about US energy policy, climate, and sustainability.

Jim DiPeso, ConservAmerica (Nov. 2012):

> One of the greatest trips of my life; I learned so much, and I really valued the opportunity to understand Germany's perspectives on energy, climate, and sustainability issues.

Curt Bramble, State Senator Utah (Nov. 2012):

> This trip was critical in understanding the German perspective on renewable energy and their transition from fossil and nuclear power. It is one thing to read and study the issue it is remarkable to see it firsthand.

Climate- and energy-focused conferences

A series of conferences hosted by the German Embassy as part of the Transatlantic Climate Bridge initiative have showcased specific areas in which Germany can offer expertise and best practices and that involve successful transatlantic cooperation. Sustainable building is one example. Germany is known in the United States as a leader in energy-efficient and zero net energy buildings and German expertise is known to have had a positive influence on the green building movement in the United States. US expertise in this field has seen major growth in recent years and there now exists a strong two-way exchange of knowledge and cooperation. The Solar Decathlon, an international green building competition for universities run by the US Department of Energy, takes place on the National Mall every two years. Team Germany won the competition in 2007 and 2009, and their winning design was used as a focus for two 'Transatlantic Climate Bridge' sustainable building conferences.

These events were hosted by the German Embassy, the German Federal Ministry of Transport, Building and Urban Affairs and Darmstadt University of Technology (winner of the Solar Decathlon 2007 and 2009), and featured a variety of German and US experts and policy makers in the energy efficiency field. Building on their success, a group of German green building experts, members of the green building community in Washington D.C., Virginia and Maryland, US government and industry representatives gathered at the German embassy in April 2009 to discuss new market opportunities through sustainable building practices. Michelle Moore, Federal Environmental Executive on Environmental Quality, gave the keynote speech and spoke of President Obama's determination to 'make sustainability a part of our mission'.

The winning house at the Solar Decathlon 2012 toured South America and North America for two years and showcased its technology in more than 20 cities. In early 2012, the waste water management operator from the city of Lingen in Germany presented its experience in regaining energy from waste water treatment. By utilizing methane captured from the anaerobic bacterial fermentation processes, 100 per cent of the plant's energy demand could be covered, saving several thousand euros each year.

During the most recent conference in October 2012, organized by the Transatlantic Climate Bridge, an expert panel presented and discussed the role of investment in the clean energy sector and its impact on green growth. The main findings during this well-attended conference in the Cannon Caucus room on Capitol Hill in Washington D.C. were that green financing is no longer a niche sector and policy has played an important role in its successful development in Germany.

The future of the transatlantic climate bridge and Germany's path toward a green economy

Since the Transatlantic Climate Bridge was launched in 2008, Germany has continued its commitment to building a green economy and setting an example for other nations' efforts to reduce their carbon footprint. This commitment has been strengthened further with the German government's 'Energy Concept' paving the way for a transformation of the country's energy policies (*Energiewende*).

The Energy Concept complements the entire package of measures taken at the European, national and regional levels. It reflects the German government's commitment to changing the energy system toward sustainability, with climate change being one main driver besides cost efficiency and energy security. The *Energiewende* enjoys major support from society and from all the political parties in Germany. In Germany, few policy issues command as much consensus as recognizing the need to fight climate change and mitigating its consequences. There are different views on exactly 'how' to do this, but not 'whether'.

The aim of the Energy Concept is not only to deal with the challenge of climate change, however. In a densely populated, industrialized country that largely relies on imports of finite commodities whose prices are constantly rising for its energy supply, it is vital to Germany's future prosperity to develop a secure energy supply at sustainable prices. In Germany, this is seen as a matter of ensuring future economic growth, increasing the capacity for technological innovation, and thus also securing jobs by securing future energy supply. Accordingly, the aim of the Energy Concept is to make Germany one of the most energy-efficient, climate-friendly and competitive economies. Above all, the programs of the energy transition are based on results of long-term studies showing clear economic, social and ecologic advantages of the low-carbon policy compared to the business-as-usual case.

Notwithstanding the long-term economic advantage to be gained by creating a green economy, Germany and its European partners know they cannot go this path alone. To counter the tendency to regulate international trade on the basis of carbon emissions, Germany believes that as many major economies as possible should move toward a low-carbon pathway. Germany therefore feels highly committed to international negotiations at the UN level, aiming at a global treaty with strong emphasis on climate protection measures.

Although climate action on a federal level in the United States may be unlikely in the coming years, there remains a commitment toward energy efficiency and the creation of green jobs in many regions throughout the United States. It is this movement that the Transatlantic Climate Bridge seeks to connect with, in the hope that Europeans and North Americans work together for common and smart energy approaches promoting low-carbon technologies and energy efficiency.

The recent boom in unconventional gas and oil resources in the US energy market has changed the once level playing field. The hope that these energy resources could pave the way for independence from energy imports dominates the discussion. In the meantime, greenhouse gas (GHG) emissions in the United States are declining due to a switch from coal to gas in the power sector, the flagging economy and slow improvements in energy efficiency. This trend masks the fact that the increased use of unconventional fossil fuels could lock in intolerable GHG emissions in the long run, which would block the United States from achieving more ambitious and responsible climate targets. At the same time, the focus on further exploitation of shale gas and oil might reduce federal and state efforts in expanding the use of renewables.

Nevertheless, with the cost efficiency of renewables and efficiency technologies continuously improving, they are seen as the major global energy future. This is also reflected in the energy program of President Obama's second term. The ongoing activities of the Transatlantic Climate Bridge will take this new situation into account, focusing on the challenges of the new energy market.

9

Formalizing the Transfer and Application of Environmental Policies and Lessons from Germany to the United States: The Case of Northern Virginia

Dale Medearis

American environmental and energy policy has tended to be insular and introspective, lacking a global perspective that is suitably tuned to regularly finding, understanding and applying lessons from pioneering countries such as Germany to the United States. It is rare to find a city, county or state agency engaged in the regular pursuit of international best practices. It is equally rare to see national urban, energy, environmental, or planning organizations engaged in formal searches, reviews and applications of lessons from abroad for application in the United States. Much of this is because most international urban environmental and energy work in the United States takes place within one of two contexts. The first context is one in which the United States exports policies, ideas and technologies to developing countries. The other context is 'soft diplomacy' and the accidental contexts of random 'social exchanges' that lack formal problem-focused, goal-oriented searches and applications of technical or policy innovations in the United States from abroad (Dolowitz and Medearis 2009).

The Asia-Pacific Partnership for Clean Development and Climate and the US-China Energy Forum are typical of the former. A review of the work of these efforts points to stereotypical development assistance projects that result in no tangible or practical economic development, energy management, climate or environmental outcomes in US cities. This is consistent with the notion that the United States stands to learn little from the rest of the world and is reflected in the research of Kingdon (1995), Lipset (1996) and Robertson (1991). In addition, the Asia-Pacific Partnership for Clean Development and Climate and the US-China Energy Forum, like too many international development assistance projects focusing on energy and climate change that involve US cities, are often viewed as patronizing or even insulting in the eyes of the developing countries, particularly given the poor performance of US cities in managing climate policies (Medearis 2012).

When efforts are made in the United States to learn about sustainable development from other countries, they are often poorly structured and fail to inform policymakers and technical staff about how the innovations overseas evolve. They also lack assessment of the performance of the foreign innovations using apples-to-apples quantitative benchmarks and often there is little prospective assessment about how pieces (rather than entire copies) of the innovations can be applied into uniquely US contexts. The lack of formal searches and testing of innovations from abroad for application in the United States has led to the marginalization of international work in general and false perception that there is nothing to be learned from countries that offer the United States valuable lessons in sustainable development. This is particularly the case at the local level.

For the past 10 years, however, Northern Virginia (as represented by the Northern Virginia Regional Commission, NVRC) has pushed forward a subtle paradigm shift in this arena. Starting with a bilateral agreement in 1999, with the Regional Planning Authority of Stuttgart *(Verband Region Stuttgart)*, the NVRC has undertaken an ambitious and long-term effort to learn from and apply lessons concerning sustainable urban development from European, but especially German cities and regions. The work with Germany emanates from the long-term history of practical and outcome-oriented transfers that have benefitted US urban development, environmental, transportation, energy and education policies since the 19th century, but weakened significantly after World War II (Dolowitz and Medearis 2009; Rodgers 1998). Building on historical precedents, NVRC's work with Germany is exceptional because of the focus and structure of searching, reviewing and testing of urban sustainability initiatives from Germany that benefit economic development, climate, energy and environmental programs in Northern Virginia. Over the past 11 years, the NVRC has conducted over two dozen peer-to-peer policy and technical exchanges with German and European regions, helped co-launch the US/German Transatlantic Climate Bridge Initiative,[1] created the first formal agreement on climate and energy between United States and European regional councils, launched the 4 Mile Run Watershed Restoration Project (a national watershed restoration plan designed with the help of German landscape architects) and created two bilateral agreements between the Commonwealth of Virginia and the German environment and transportation ministries.

The work with Germany has reshaped local environmental, transportation, energy and climate mitigation planning in Northern Virginia. In 2009, the NVRC developed and started implementing the first comprehensive Community Energy Plan for Loudoun and Arlington counties – now recognized by the National Association of Counties as national models. Similar-scale plans are now underway in other parts of Northern Virginia. These plans have been very much informed by the energy and urban development policies of Germany – from district energy, to solar PV to transportation design and management.

This chapter highlights the evolution of Northern Virginia's work to develop and sustain a problem-focused and goal-oriented search and testing of innovative sustainable development policies from Germany to the United States. Special emphasis will be given to the evolution of cross-national policy transfers characterized by random searches and informal structures, into a more rational and formal process characterized by problem-focused and goal-oriented searches for information, review, debate and testing of that information. In addition, to help frame the context of Northern Virginia's story, this chapter reviews some of the national-level history and precedents with transferring urban environmental and sustainability policies from Germany to the United States.

Problems with cross-national policy transfers in the United States

A critical variable in the cross-national policy transfer equation that commonly slows formal efforts to find and apply lessons, especially from European countries such as Germany, to the United States appears to be the phenomenon of American exceptionalism. The notion that the United States stands to learn little from the rest of the world can be traced to research by Kingdon (1995), Lipset (1996) and Robertson (1991). These authors have suggested that within the United States, there is a prevalence of anti-state, individualist and anti-egalitarian attitudes on which policymakers rely to justify not engaging in formal searches and evaluations of innovations from abroad. Dolowitz and Medearis (2009) find multiple signs of this phenomenon in their research about urban environmental policy transfers from Germany to the United States.

In Northern Virginia, the regional council of governments (the Northern Virginia Regional Commission) has bucked this trend of aimless, event-reliant, cross-national exchanges that has plagued past international cooperation at the sub-national level. The NVRC has created purposeful and structured transfers and applications of multiple regional and urban development innovations from Germany. The work has consisted of formal searches, reviews and tests of policy innovations from Germany and their suitable inclusion into Virginia. The purposeful study and analysis has led to a successful cycle of outcomes and tangible results among the local authorities in the region that has strengthened and expanded acceptance of learning from abroad.

Northern Virginia: Background

Northern Virginia is a diverse region demographically, geographically and economically. The region is home to approximately 2.3 million people spread out over more than 2,000 square kilometers – including 150 kilometers of coastline within the Chesapeake Bay watershed. According to the George Mason Center for Regional Analysis, Northern Virginia represents approximately half of the $313 billion gross regional product of the Metropolitan Washington region (George Mason University 2008). The region is strongly influenced by the US government, with one-third of the regional economy tied to federal employment or related contracting and services. However, over two-thirds of the regional economy emanates from international, national and local businesses – including a diverse collection of globally interconnected information and Internet-related industries (George Mason University 2008).

Governmentally, Northern Virginia is composed of 14 local authorities brought together under the auspices of the NVRC, a political entity of the Commonwealth of Virginia established to encourage and facilitate local government cooperation on a range of regional climate, energy, environmental and social issues. The NVRC is led by a 25-member Board of Commissioners composed of elected officials appointed by the governing bodies from the 14 local authorities.

Environmental and energy planning challenges in Northern Virginia

Over the past decade, Northern Virginia has been confronted by a range of serious energy and environmental challenges, particularly the issues of climate change. In addition, the metropolitan Washington DC region is one of the most rapidly growing regions in the United States and must plan for the inclusion of more than 1.5 million new residents in the next 20 years. This will place enormous demands on housing, mobility and energy. The Metropolitan Washington Council of Governments (MWCOG 2008) estimates energy consumption in the metro Washington region will rise by 40 percent by 2050. The region is already a significant consumer of energy as compared to the rest of the world. According to the US Energy Information Administration, annual per capita energy consumption in the Commonwealth of Virginia is 345 million BTUs. By comparison, Germany consumes 176 million BTUs per capita annually and France, 182 million BTUs.[2] Moreover, a recent energy study for one of our largest counties indicated that annual energy use per square meter of residential and commercial space was about 230,000 BTUs per square foot, a level at least twice that of the EU for comparable services and climate (Medearis, Garforth and Bluem 2010).

Increased energy consumption and demand will strain the region's energy infrastructure, electricity grid and gas and transportation networks. It will pose threats to economic stability and prosperity if energy prices rise as they did between 2000 and 2005, when prices increased 14 percent for electricity, 53 percent for natural gas, and 68 percent for gasoline. Switching to current renewable energy options alone is not a viable solution to reducing greenhouse gas (GHG) emissions because of the increasing demand for energy.

There are similar opportunities and challenges with respect to buildings due to the rate of growth in this area. In the United States, there are currently 20,000 registered buildings under the LEED rating system of the US Green Buildings Council (USGBC). However, the rate of certifying new buildings and retrofitting existing buildings would have to increase exponentially to cover a significant portion of the approximately 129 million homes and 10 million commercial buildings in the United States within a reasonable time. There is strong evidence that simply relying on voluntary rating systems, such as LEED, to create sustained high levels of energy efficiency may not be effective. In addition, many communities of the region aspire to reduce greenhouse gas emissions 80 percent by 2050, but struggle with the proper policy responses that will have the desired effect. The actions to date by most municipalities are confined to government activities and buildings, which often represent less than 5 percent of GHG emissions. The deployment of solar PV is currently less than 1 MW; there are no more than two passive houses in the region. Clearly, there is a paradigm shift waiting to occur beyond business as usual (Garforth and Medearis 2011).

In addition, the population growth is anticipated to drive up emissions of greenhouse gases and exacerbate the region's current non-attainment status vis-à-vis federal air quality standards, particularly ozone and particulate matter. The Brookings Institution quantified transportation and residential GHGs for the 100 largest US metropolitan regions. The Washington-Arlington-Alexandria-DC region ranked 100, Baltimore-Towson ranked 91 and Virginia Beach-Norfolk ranked 77 (Brookings Institute 2008). These poor results are largely due to the reliance on coal-based electricity, car-dependent transportation systems and inefficient homes and buildings.

In the realm of climate adaptation, the existing science and data suggest that the climate in Northern Virginia has changed and that its effects are already being felt. Average mean air temperatures in the region have risen 2 degrees Celsius since 1970. The warming trends suggest an increase in severe storm events and sea-level rise between 10 and 25 centimeters in Mid-Atlantic and Chesapeake Bay regions (Virginia Energy Plan 2007, p. 166; National Wildlife Federation 2008, p. 4). The potential rise of the Chesapeake Bay and intensity of storm events are expected to lead to sustained flooding and property loss for coastal communities in Northern Virginia such as Alexandria and Fairfax County. The rise of temperatures and sea-level also have caused species displacement and are likely to further stress the region's water infrastructure systems, particularly storm water (MWCOG 2008).

Historical precedents of transferring lessons from Germany to the United States

There is an abundance of historical literature recording anecdotes and outcomes of urban and environmental transfers from Germany to the United States. The available literature, particularly for the era 1870–1945, indicates a number of relatively active, problem-focused searches for lessons from Germany, particularly in the fields of education, urban planning and natural resource management. The time period between 1870 and 1945 was when 'the reconstruction of American social politics was part of a movement of politics and ideas throughout the North Atlantic world that trade and capitalism had tied together' (Rodgers 1998, p. 3). Rodgers (1998, pp. 6–7) adds that the development of reform policies in the United States was started by 'a sudden abundance of solutions, a vast number of them brought over through the Atlantic connection'. Information about an innovation or lesson traveled to the United States through networks of interconnected political and academic elites. These forces worked to transform the US university system and management of US natural resource policies.

Patterns of the voluntary transfer of urban planning, education and other social policies during the late 19th century followed what could be interpreted as relatively structured processes. Signs emerge from the literature that laggards (the United States) followed the pioneering models of German academia and resource management not only out of desire to appear modern, but because of relatively informed and structured searches for information and efforts to apply that information in the United States. In the late 19th century, Germany seemed a world leader in education, environmental and urban planning policies (Rodgers 1998, p. 4). American urban planners, academics and conservationists confronted challenges, heard from third parties about innovative ideas and solutions in Germany and endeavored to learn and apply them in the United States. They committed time, money and other resources for travel to Germany to gather understanding and knowledge about innovative programs (Rodgers 1998; Miller 2001; Brubacher and Rudy 1997).

Brubacher and Rudy (1997) relate how the entire US academic system, including the development of the modern research university and doctoral programs was modeled after pioneering academic systems in Germany. They add that '[t]he impact of German university scholarship upon nineteenth-century American higher education is one of the most significant themes in modern intellectual history'. Brubacher and Rudy (1997, p. 174) describe how the critical concepts of the German university system, such as scientific research through original investigation or the specialists' lecture, were imported to institutions such as the University of Michigan and Johns Hopkins University. Driven by concerns about poor standards in the American university system, over 'ten thousand American students passed through the halls of Germany universities between 1815 and 1914' (Brubacher and Rudy 1997, p. 175). With respect to outcomes, Brubacher and Rudy (1997) point to efforts by Daniel Gilman, the president of Johns Hopkins University, to avoid duplication of existing models in Germany. Rather, Gilman pursued an effort to 'supply the needs of the United States in certain specialist learned fields, such as language, mathematics, ethics, history and science' (Brubacher and Rudy 1997, p. 178).

The period between 1900 and 1930 is considered to be the start of contemporary urban planning in the United States (Dumpelmann 2005). By coincidence, the period was considered the 'rationalistic era' because of planners' emphasis on hygienic and social functions of parks (Dumpelmann 2005). The era was characterized by similar efforts of American academicians and the transfer of urban planning lessons from Germany. 'Grand Tours' to European capitals were organized to study park designs, transportation planning and taxation policies (Sutcliffe 1981). Moved by the crisis of cholera outbreaks in New York City, Benjamin Marsh deliberately moved to Germany to understand city planning practices that emphasized human health and hygiene (Peterson 2003; Rodgers 1998). Marsh considered German urban planning systems in general, but the concept of zoning in particular, to be a model for public health planning practices for the United States. In 1901, Marsh traveled to Germany and studied the German language and Frankfurt-am-Main's planning codes. Marsh returned to New York City to introduce the city's first comprehensive zoning regulations (Sutcliffe 1981).

Other noticeable transfer efforts in the realm of urban planning included Senator John MacMillan's work to transform the National Mall in Washington D.C., by finding and applying positive urban park and streetscape policies from Europe. The tours by the MacMillan Commission to Berlin and other European capitals in 1901 profoundly shaped the eventual design of the National Mall (Sutcliffe 1981). Returning from the MacMillan Commission tour of Europe, Frederick Law Olmsted, Jr., testified before the US Congress in 1906 that German urban planning practices offered important lessons (and mistakes) for planning efforts in the United States. Although Great Britain was also considered a model for Canadian and American planners (Ward 1999), the efforts by so many American planners, environmentalists, scientists and education specialists to go to Germany to learn, despite poor communication networks and language and cultural barriers, suggests a relatively thoughtful transfer process.

In the realm of natural resource management and planning, Gifford Pinchot studied natural resource and forestry management practices for 12 months in Bavaria, Germany, in 1888, before returning to the United States to create the US National Forest Service. Miller (2001) records how a range of German forestry harvesting and maintenance practices were emulated by Pinchot after Bavarian models. Aldo Leopold traveled to Germany in 1936, to better understand resource conservation laws and policies. After six months in Germany, Leopold returned to Wisconsin to develop the state's first resource conservation programs (Leopold 1936).

Rodgers (1998) writes that following World War II, the United States became a world hegemon and that the 'entrance of the United States onto the international political stage was also an exit – the advent of the American century' (Rodgers 1998, p. 488). The era represented a time when comparisons of conditions in other countries were not as useful a justification to look abroad for lessons. As Rose (1993, p. 34) observes, it could be considered a time when 'national power was used to afford not to learn' in the United States.

The emergence of a new American century corresponded with the evolution of 'exceptionalist' attitudes toward the importation of environmental and planning ideas after World War II. This was particularly noticeable in the realm of land-use planning. Kayden (2000) describes how anti-state attitudes and individualism in the United States preclude the transfer of European-style national land-use planning policies. These notions are touched on by Lefcoe (1979) and Beatley (2000). Bruegmann (2005) has identified population densities and demographic stagnation in Germany as important differences affecting adoption of land-use planning models in the United States.

Northern Virginia's 'exceptionalism'

Northern Virginia's efforts to systemically find, understand and test lessons from Germany are unique and reflect an approach to cross-national policy transfer that deviate from the conventional route of copying content of policies and assuming that application will occur automatically. Instead, the NVRC has started to work with cross-national policy transfer by tying the focus on understanding policy content with the process of evaluating prospectively which relevant pieces of an imported innovation potentially fit into the unique Northern Virginia policy landscape. This represents a substantial deviation from normal international cooperation among cities, regions and states in the United States.

The NVRC's international work with Germany started in early 1999, when it looked to establish a formal partnership with a European region that had attributes such as innovative spatial and environmental planning, governance, transportation policies and equivalent geographic and economic features. The NVRC observed and studied the work of the Greater London Authority, the regional councils of Stuttgart and Hannover (Germany) and the regional council of Copenhagen. After review and consideration of the options, the NVRC selected the Verband Region Stuttgart because of the relatively similar geographic and population sizes of the regions, its innovative landscape plan and the Stuttgart region's long-term history of working with the United States.

The Stuttgart region was then and is still regarded internationally as a leader in regional spatial planning and 'green' infrastructure. Ann Spirn's *Granite Garden* (1985) profiled the Stuttgart region's experiences with regional landscape planning, and particularly the phenomenon of 'clean air corridors'. Spirn revealed the extent to which the Stuttgart region modeled its green infrastructure's air quality benefits, and how those benefits were factored into land-use planning and even building designs for the region. Spirn's work itself was informed by other prominent research that profiled the utility of drawing lessons from Germany to the Northern Virginia region, including Ian McHarg's 'Potomac River Basin Study of 1965–1966'. In this seminal work, McHarg highlighted how the ecological restoration of the Potomac could be informed by practices with holistic watershed restoration and waterfront development in Hamburg (and Amsterdam and Stockholm).

In 2000, after two individual peer-to-peer policy exchanges, a formal five-year agreement of cooperation was developed and approved between the NVRC and the Verband Region Stuttgart. In the summer of 2000, members of the NVRC board and secretariat traveled to Stuttgart to observe and understand Stuttgart's regional planning policies, including its work with green infrastructure. The NVRC staff and board members studied Stuttgart's green infrastructure policies, the design of clean air corridors, and their air quality effects. Two summers later, in 2002, storm water and green infrastructure experts from Northern Virginia took part in a second peer-to-peer policy exchange to Schleswig Holstein, the Ruhr Valley and Stuttgart to explore further applications of storm water management to Northern Virginia and the Potomac River watershed.

Subsequent to the exchange, the NVRC co-developed a one-day 'green infrastructure' workshop at the German Embassy to explore applications of the low-impact development and green infrastructure lessons from Germany to the Potomac River watershed. The workshop tightened the focus on specific design and policy reforms in Virginia concerning low-impact development stormwater practices. These changes were ensconced into the 2004 launch of the Four Mile Run comprehensive watershed restoration plan, which involved prominent European landscape architects who had worked on similar waterscape plans in the Stuttgart region. Parallel to this effort, NVRC staff actively informed the development of the first comprehensive set of state-wide technical guidelines for low-impact storm water management in Virginia.

Between 2003 and 2008, several more peer-to-peer policy exchanges were conducted between practitioners from Northern Virginia and Germany. The themes of the exchanges included regional and transportation planning and emerged with several outputs. Real-time signage was introduced to the Washington DC metro system, as well as traffic-calming projects in Fairfax County. In addition, 'green' roofs started to proliferate across the Washington DC region. It has been reported that the

Washington DC region has among the highest concentrations of green roofs in the United States.

In 2008, the Northern Virginia Regional Commission expanded its bilateral work with Stuttgart to include the 70 largest regional councils in Europe by convening the first meeting of US and European regional councils around the theme of climate change. Working with the National Association of Counties and the European counterpart for regional councils, METREX, NVRC developed a 'Declaration of Cooperation' on climate and energy. In the same year, the NVRC co-initiated the Transatlantic Climate Bridge in Berlin, and developed a bilateral agreement between the Commonwealth of Virginia and the German Federal Ministry of Environment on the exchange and application of best practices in energy and climate policies.

The additional paradigm shift in transatlantic cooperation occurred in 2009, when the NVRC in partnership with the German embassy, launched the first of two Community Energy Plans (CEPs) for the Northern Virginia region. CEPs are long-term, quantitatively informed and benchmarked master plans for greenhouse gas mitigation and energy management for local governments. CEPs are plans that successfully frame energy security, and environmental challenges by integrating energy efficiency, heat recovery, renewable energies, energy distribution, transportation alternatives and sustainable land-use development. German cities such as Hamburg, Freiburg and Stuttgart embody the practice of CEP, as evidenced in the thoughtful integration of district energy, renewable energy, robust standards for energy efficiency in buildings and integrated land-use and transportation policies.

In July, 2009, Loudoun County and the Northern Virginia Regional Commission worked to develop the first CEP in Northern Virginia. The 2009 Loudoun Energy Strategy is now recognized as a model for local energy planning in the United States. It is also a model for transatlantic cooperation. Among the first in the United States, a 40-year energy and climate mitigation plan, benchmarked against German cities, with quantitative modeling and indicators calculated by German and European consultants, was developed and approved. In 2010, the National Association of Counties recognized the Loudoun CEP as a national model for local energy and climate planning. In 2011, Arlington County completed a more robust planning process with the help of an equivalent European and NVRC-led team.

The focus on heat recovery and cogeneration in the urban setting are relatively underemphasized elements, representing one of the substantive differences between the United States and German approaches to CEP planning. To overcome some of the cultural, technical and legal barriers that often are used to eliminate district energy from consideration in local energy planning in the United States, the NVRC commissioned a legal paper to profile how within existing Virginia law, district energy systems from German cities such as Stuttgart, Mannheim and Berlin could be applied to Northern Virginia. The report 'District Energy Systems: An Analysis of Virginia Law' (2010) represents a rare precedent in the transatlantic conversation and the creation of a very specific, problem-focused and goal-oriented approach to addressing cross-national policy transfer of district energy systems.

In 2011, the cooperation between the NVRC and German regions such as Stuttgart and the German federal government deepened. The NVRC drafted a Declaration of Intent of Cooperation between the Common wealth of Virginia's Department of Transportation, and the German Federal Ministry of Transportation, Housing and Buildings. In addition, the NVRC launched the 'Transatlantic Urban Climate Dialogue' together with the Freie Universität Berlin's Environmental Policy Research Center. This two-year project will further expand the review and analysis of transatlantic energy and climate policy learning and between three German and two North American urban metropolitan areas.

Conclusion

As US cities and regions work to develop innovative low-carbon, high-quality economic development and efficient affordable energy policies, they will need all the tools available. Drawing from successful experiences of pioneering countries such as Germany will be an important potential source of competitive advantage. But the transfer of these innovations does not occur automatically. It requires investment and study within problem-focused, goal-oriented and geographically specific contexts. For the past decade, the Northern Virginia Regional Commission has worked successfully to be the US model in this context.

Notes

1. See the chapter by Kueppers-McKinnon, Maue and Kristan in this volume.
2. The gap is even wider since the Virginia numbers do not include national defense, air traffic, maritime and national industrial emissions.

References

Beatley, Tim (2000) *Green Urbanism: Learning from European Cities* (Washington, DC: Island Press).

Brookings Institute (2008) 'Shrinking the Carbon Footprint of Metropolitan America'. Retrieved from http://www.brookings.edu/~/media/Files/rc/papers/2008/05_carbon_footprint_sarzynski/carbonfootprint_brief.pdf.

Brubacher, John and Willis Rudy (1997) *Higher Education in Transition: A History of American Colleges and Universities* (New Brunswick, NJ: Transaction Publishers).

Bruegmann, Robert (2005) *Sprawl: A Compact History* (Chicago, IL: The University of Chicago Press).

Dolowitz, David and Dale Medearis (2009), 'Considerations of the obstacles and opportunities to formalizing cross-national policy transfer to the United States: A case study about the transfer of urban environmental and planning policies from Germany', *Environment and Planning*, 27/4, 684–697.

Dumpelmann, Sonja (Fall 2005) 'The Park International: Park System Planning as an International Phenomena at the Beginning of the Twentieth Century' (Bulletin of the German Historical Institute, Washington, DC).

Ewing, Reid (2007) 'Growing Cooler: The Evidence on Urban Development and Climate Change' (Urban Land Institute, Washington, DC).

Garforth, Peter and Medearis, Dale (2011), 'Green Jobs Strategy and the Transition to a Low Carbon Economy', OECD Local Economic and Employment Development (LEED) Working Papers, No. 2011/07, OECD Publishing. DOI: 10.1787/20794797.

George Mason University (2008) 'Washington Area Economic Outlook'. Retrieved from http://www.cra-gmu.org/forecastreports/08forecasts/EconomicOutlook-GFOA.pdf.

Greater Washington Board of Trade (2007) 'Key Findings and Green Benchmarks: A Report from the 2007 Potomac Conference' (Washington, DC).

Kayden, Jerold S. (2000) 'National land-use planning in America: Something whose time has never come', *Washington University Journal of Law and Policy* 3, 445–472.

Kingdon, John W. (1995) *Agendas, Alternatives, and Public Policies*, 2nd ed. (New York: Harper Collins).

Lefcoe, George (1979) 'The right to develop land: The German and Dutch experience', in George Lefcoe (ed.), *Land Development in Crowded Places: Lessons from Abroad* (The Conservation Foundation, Naperville, Illinois).

Leopold, Aldo (1936) 'Naturschutz in Germany', *Bird Lore* 3/2, 102–111.

'Lipset, Seymour M. (1996) *American Exceptionalism: A Double-Edged Sword* (New York: WW Norton Company).

Medearis, Dale (2012) 'Northern Virginia region applies global lessons', League of Cities. Retrieved from http://www.nationscities-digital.org/nationscities/20120326?pg=6#pg6, accessed 15 March 2013.

Medearis, Dale and Peter Garforth (2011) 'Green jobs strategy and the transition to the low carbon economy in Northern Virginia' (Organization for Economic Cooperation and Development, Paris).

Medearis, Dale, P. Garforth and S. Bluem (2010) 'Promoting Energy Innovation and Investment Through Transatlantic Transfer of Community Energy Planning Policies', Policy Paper #43, The American Institute for Contemporary German Studies (Washington, DC).

Metropolitan Washington Council of Governments (2008) 'Draft Climate Change Report and Recommendations'. Retrieved from http://www.mwcog.org/uploads/committee-documents/a15eXlZY20080523134421.pdf, accessed 11 March 2012.

Miller, Char (2001) *Gifford Pinchot and the Making of Modern Environmentalism* (London: Island Press).

National Wildlife Federation (2008) 'Sea-Level Rise and Coastal Habitats of the Chesapeake Bay' (Reston: National Wildlife Federation). Retrieved from https://www.nwf.org/~/media/PDFs/Global-Warming/Reports/NWF_ChesapeakeReportFINAL.pdf?dmc=1&ts=20130314T1310037031, accessed 15 March 2013.

Peterson, Jon A. (2003) *The Birth of City Planning in the United States, 1840–1917* (Baltimore, MD: Johns Hopkins University Press).

Pollin, Robert (2008) 'Job Opportunities for the Green Economy: A State-by-State Picture of Occupations that Gain from Green Investment' (University of Massachusetts, Amherst). Retrieved from http://www.peri.umass.edu/Green_Jobs_PERI.pdf.

Robertson, David B. (1991) 'Political conflict and lesson drawing', *Journal of Public Policy*, 11/1, 55–78.

Rodgers, Daniel (1998) *Atlantic Crossings: Social Politics in a Progressive Age* (Cambridge, Massachusetts: Harvard University Press).

Rose, Richard (1993) *Lesson Drawing in Public Policy: A Guide to Learning Across Time and Space* (Chatham, NJ: Chatham House).

Spirn, Ann (1984) *The Granite Garden* (New York: Basic Books).

Sutcliffe, Anthony (1981) *Towards the Planned City: Germany, Britain, the United States and France 1780–1914* (New York: St Martin's Press).

United States Senate. Senate Document No. 422, 61st Congress, 2nd Session. Washington: Government Printing Office, 1910. 'City Planning. Hearing Before the Committee on the District of Columbia, United States Senate on the Subject of City Planning'.

Virginia Department of Mines, Minerals and Energy (2007) *The Virginia Energy Plan* (Richmond, Virginia), (Department of Mines, Minerals and Energy, Virginia State Government, Richmond Virginia). Retrieved from http://www.dls.virginia.gov/groups/energy/VEP.pdf.

Virginia Department of Mines, Minerals and Energy (2008) 'Commodity Fact Sheet'. Retrieved from http://www.dmme.virginia.gov/DMR3/energyresources.shtml.

Ward, Stephen V. (1999) 'The international diffusion of planning: A review and a Canadian case study', *International Planning Studies* (4), 53–78.

10
Cultivating the Cross-Cultural Engineer: Key Insights

Dana Elzey and Kerstin Steitz

Introduction

The central thesis of this chapter is that intensive, short-term study abroad programs, designed for engineering students can provide significant benefits for professional development. Among the most important of these is the recognition that engineering is fundamentally a cultural activity. How engineers are educated and trained, the professional practice of engineering and the engineering design process are all culturally mediated. This leads to increased emphasis on intercultural awareness and enhancement of specific skills essential for professional engineers in a globalized environment. These skills are brought strongly into play for engineers engaged in open-ended problem solving (design thinking) in a diverse, multicultural team setting. Study abroad programs for engineers and other professionals can be made more effective in developing these skills by directly engaging participants in cross-cultural design thinking.

We describe two short, study abroad programs, developed and implemented at the University of Virginia (UVA), followed by seven key insights gained by participants in the programs. We draw our insights and conclusions regarding learning outcomes from various sources: final papers, presentations and design projects, student design portfolios, students' journals, reflections during the course of the study abroad, as well as reflective essay questions. Both study abroad programs, *Global Technology Practice (GTP)* and *Global Ingenuity 21 (GI 21)*, are two weeks in duration, are offered during the summer, provide three credit hours of technical elective, and take place in Germany. German university partners play a critical role in the planning and implementation of both programs, but are accompanied by a UVA engineering school faculty program director throughout. The program director guides the learning process through group discussion with practicing engineers and technology managers during site visits, and frequent, 'off-line' reflective group discussion. Although the approach and character of the two programs are quite different, they share a number of key learning outcomes.

A further aim of this article is to emphasize that programs for engineering students, which incorporate opportunities for cross-cultural design thinking, can be powerfully efficient in helping students acquire many of the competencies established by the Accreditation Board for Engineering and Technology (ABET 2000). Among other skills, ABET 2000 requires engineering programs to demonstrate that their graduates have '(c) an ability to design a system, component, or process to meet desired needs, (d) an ability to function on multi-disciplinary teams, (e) an ability to identify, formulate, and solve engineering problems, (f) an understanding of professional and ethical responsibility, (g) an ability to communicate effectively, (h) the broad education necessary to understand the impact of engineering solutions in a global and societal context, (i) a recognition of the need for, and an ability to engage in life-long learning, (j) a knowledge of contemporary issues'.[1] Despite their brevity, short, intensive study abroad programs integrating creative problem solving in a culturally diverse setting can greatly advance engineering students' progress toward these learning outcomes.

Global Technology Practice: 'How the engineer studies, works and lives in Germany' (Stuttgart, Germany)

Engineering students are seldom offered the opportunity to view the education, training, professional practice and social life of engineers in a given culture, even their own. This is what the Global Technology Practice (GTP) program, a two-week, summer study abroad program in Stuttgart, Germany does. Through site visits to basic and applied research institutions, industry and the German Society of Engineers, as well as participation in various cultural activities, the GTP program allows participants to see the making of the German engineer and the factors that contribute to Germany's reputation for excellence in engineering. They also develop a better appreciation for the influence of cultural values, tradition and history on how engineering problems are prioritized and the approach taken to address them. An inevitable, and very valuable, by-product of this experience is the comparison and contrast of German engineering with what they know and understand (or assume) about engineering and engineers in the United States.

Figure 10.1 illustrates schematically, the structure of the GTP program. Although listed in the course catalog as an engineering course (ENGR mnemonic), the course is open to students outside engineering; no specialized knowledge or prerequisite coursework are required. There is also no language requirement. Basics, such as greetings, being polite, ordering food in a restaurant, asking for help, etc., are integrated into a series of pre-departure meetings. The course instructor/program director lives, travels and works with the participants throughout the program. In addition to maintaining a daily journal, group discussions at the end of each day provide a framework for guided reflection and assimilation of new knowledge and insights. The cornerstone assignment for the three-credit course is a research paper on a topic selected by the student prior to departure. The central requirement for an acceptable research topic is that the planned site visits are deemed likely to allow the student to access information and insights relevant to the topic. Site visits, cultural experiences, whether sitting in a German *Biergarten* or hiking in the Alps, conversations with German engineering student peers, etc. are all considered data collection opportunities in a living laboratory. Industry site visits, such as Bosch, Mercedes Benz, Siemens, etc., focus on opportunities for discussion with engineers and technology managers, and typically include, but do not rely, only on a tour of the manufacturing facility. Students are encouraged to ask questions about the speaker's educational and professional path, their everyday experience as an engineer and their aspirations and plans for the future.

Technology areas represented most strongly by the GTP itineraries over the past five years are listed in Figure 10.1. Students are perhaps most strongly affected by their experiences and observations related to sustainability and renewable energy. It becomes quite clear that these are not merely areas of advanced technology, as many of them may have presumed, but are in fact, reflections of deeply held cultural values and attitudes, e.g. toward nature. One's ability to develop and implement a sustainable infrastructure is conditioned by a society's view of the relative importance of the challenges it faces, which in turn rests upon cultural values, historical experience and other non-technological factors.

The GTP program is hosted by the Universität Stuttgart, as part of an asymmetric exchange partnership with the University of Virginia's School of Engineering and Applied Science (SEAS). The exchange balance is accounted for in terms of student-weeks; 15 UVA participants in Germany for two weeks equals 30 student-weeks, enabling one Stuttgart student to attend the School of Engineering one year (30 weeks) tuition-free, or two German engineering students for one semester each.

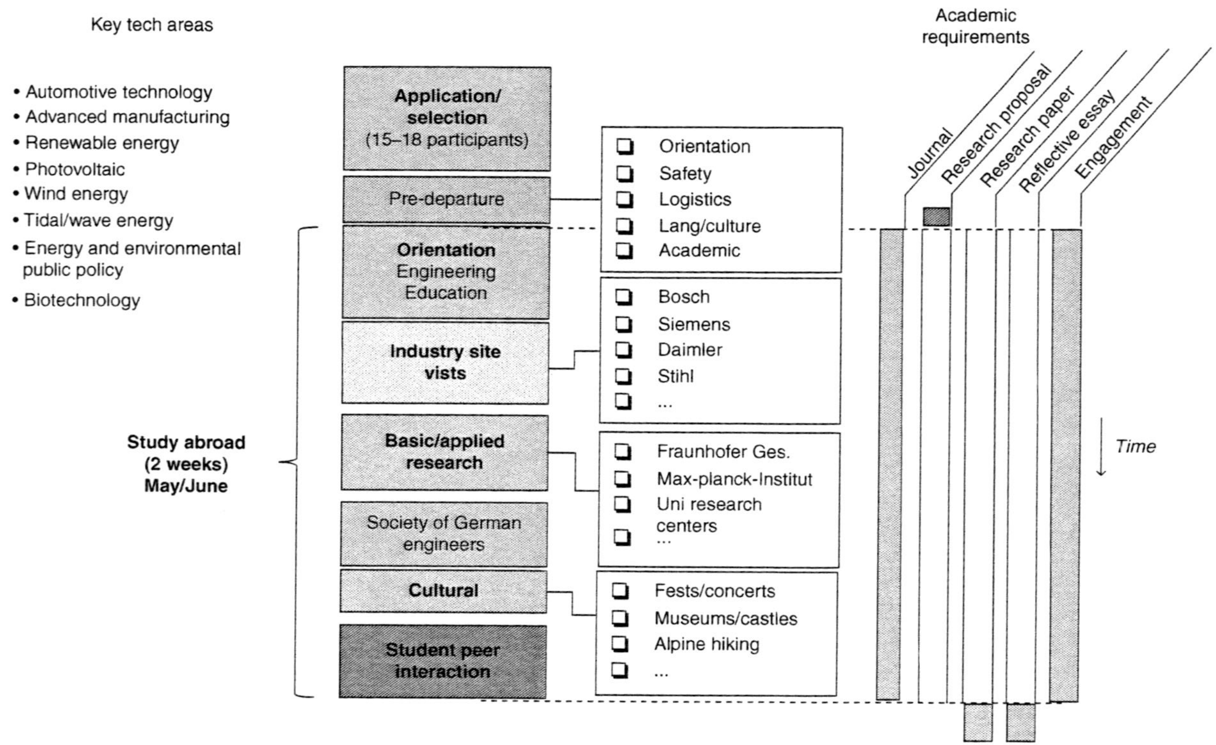

Figure 10.1 Schematic illustration of the structure and academic requirements of the GTP study abroad program in Stuttgart

In addition to gaining knowledge about education and engineering practice in Germany, participants, all of whom are undergraduate students, obtain a fuller understanding of their own educational and career path. Having made the observation, for example, that industry and academia work together in the education and training of the engineer differently in Germany than in the United States, one asks 'why?'. Over the five years the GTP program has been in existence, there have been a number of recurring observations as to deeper insights acquired, changes in awareness and attitude, development of new skills, etc. Much of this information has been obtained through students' journal reflections, reflective essay questions, research papers and debriefing, following the course. Some of the more significant observations are described more fully below.

Global Ingenuity 21: A cross-cultural engineering design 'think tank' (Braunschweig, Dresden, Berlin)

The Global Ingenuity 21 program is an intensive, two-week study abroad program in Germany for undergraduate students in the School of Engineering and Applied Science (SEAS) at UVA, and engineering students at the Technische Universität Braunschweig. Taking place in early summer (for the first time in 2010), the 15–18 UVA students spend a total of 14 days in Braunschweig, Dresden, and Berlin. The core of the program is a 9-day, cross-cultural engineering design 'think tank' experience, in which the UVA students, together with TU Braunschweig engineering students, seek a creative, conceptual solution to a design challenge (see Figure 10.2). The design challenge is put forward by the program's sponsor, Volkswagen Group of America (VWGoA). The efficient and effective search for innovative solutions to open-ended (design) problems constitutes one of the foremost critical skills for the engineer of the 21st century. Experiences in which students develop and exercise design-thinking skills while working across cultural boundaries are still rare, but of enormous potential value. Engineers in particular, are increasingly asked to work across cultural boundaries, or are engaged in the design of products and services for international markets.[2]

There are several important, and to some extent unique, considerations necessary for the operation of the GI 21 Program. While the UVA group is diverse with regard to year and major, the TU Braunschweig group is more homogenous, deriving for the most part from a single department. Another challenge is that the TU Braunschweig students take the course in addition to their regular course work. Thus, they are

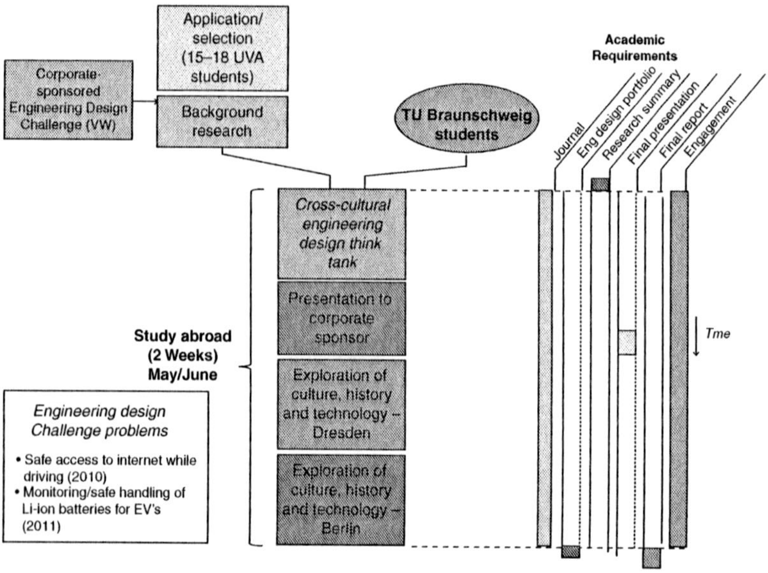

Figure 10.2 Schematic illustration of the structure and academic requirements of the Global Ingenuity study abroad program in Braunschweig, Dresden and Berlin

not able to attend all working sessions and site visits. In addition, the UVA group tends to outnumber the TU Braunschweig students roughly 2-to-1. These considerations affect the dynamics of teamwork, roles and attitudes within the project, and therefore require the attention of program facilitators.

Another important factor is the choice of the design challenge. A successful problem statement is one that combines ample opportunity for creative thinking, an acceptably low level of specialized expertise, and opportunities to apply design and analytical knowledge and skills. It is also helpful if the design challenge is relevant to the interests and capabilities of VW and other R&D and manufacturing facilities, local to the Braunschweig area, which can then become the subject of site visits. The student team should also have some flexibility in developing alternative interpretations of the challenge. One student remarked in the final course evaluation on how surprising, yet refreshing, it was to have the freedom to redefine the problem, since this was something that would probably never happen in the classroom. Examples of design challenges included the development of a means to enable safe driving while accessing electronic devices, such as phones, GPS and the Internet, or of a safe means of recycling lithium-ion batteries.

Preparation prior to the cross-cultural think tank is essential to the success of the program. The UVA student group receives the client statement of the problem in January and spends the next three months, or so, carrying out background research on the problem, its context (socio-cultural, economic, political, environmental, etc.), the prior art and current state of-the-art. The group meets every 2–3 weeks to discuss findings, assess progress and assign new responsibilities. It is also during this phase that the group sharpens its perception of the 'real' problem, and a sense for the sub-problems and areas of expertise needed emerges.

At the start of the think tank, the cross-cultural team usually has differing views of what the actual problem is, and sets differing priorities among design requirements. For example, when discussing the distracted driving problem, German students were generally more skeptical as to why anyone would want to access the Internet while driving. On the other hand, American students easily progressed from checking their email while driving, to examining other useful purposes of the Internet in the car, such as the delivery of customized geospatial information.

At TU Braunschweig, participants spend roughly 5–8 hours per day working in the cross-cultural think tank. Although the program director is almost always present during the think tank, participants lead and conduct the work themselves. In addition to their own research and design work, they attend lectures and make site visits to various centers for research and engineering, the VW facilities and manufacturing plant in Wolfsburg, the AutoUni, etc. With time abroad so limited, the arrangement of guest lectures and site visits need to be scheduled to build on each other, while leaving enough time for questions, spontaneity and creativity. It is very important to allow for enough time after or during each lecture/site visit for students to ask questions and reflect upon what they have just learned. As someone who may be very familiar with the culture and site within the host country, the study abroad program director can easily underestimate how new and overwhelming everything is for the inexperienced program participant. It is therefore wise to avoid scheduling too many events for one day.

At the conclusion of the think tank portion of the program, the cross-cultural team presents its final solution concept to VW managers, engineers and other interested attendees at the AutoUni in Wolfsburg. Following the final presentation, the UVA group travels to Dresden (three days) and Berlin (three days) to learn more about Germany's culture and history. As a result of the experience of working in cross-cultural teams and solving a design problem, the students are much less likely to think of themselves as tourists during these trips, but rather as researchers or as global (cross-cultural) engineers. The insights we gained from their reflective journals, their engineering design portfolios, as well as their responses to the reflective essay questions, indicate that they independently continue the process of cross-cultural learning during these visits. The differences between Germany and the United States, whether with regard to public transportation or recycling, are explored and considered as culturally mediated solutions arising from different views of the 'real' problem, or of how to prioritize design requirements.

Significant learning/insights

Real problems are best addressed in multi-disciplinary centers for R&D

Students participating in the GI21 and GTP programs in Germany have often been struck by the seemingly odd organization of German universities. Firstly, the extent to which industry is present in academia, in the form of physical facilities, financial support of basic and applied research and development and in determining the nature of the research and design work being done, is surprising. Undergraduate students in German universities are much more likely to be found working to solve a practical challenge faced by a particular industry. For example, they might be using 3-D modeling software to optimize flow during pressure casting of a crankcase housing for an internal combustion engine, or developing software to collect and analyze specific sensor data for an autonomous vehicle project. This is not to say that students aren't doing this sort of work at US universities and engineering schools, but that the prevalence of this type of project and the closeness of collaboration between industry client and academia is greater. This is relevant to the now familiar discussion of the 'engineering science' model favored by US engineering schools since the 1950s, which focuses on theory while neglecting the practice of engineering.[3]

The second surprising observation is that German universities are often organized, not around disciplines, but around problems. While there are certainly advantages to having all faculty and students with interest and expertise in mechanical engineering work in the same building, it is also clear that real problems are typically not solved by people knowledgeable in any single discipline. Even apparently narrowly defined problems, such as the development of more impact resistant bumpers, benefit from the collaboration of say, materials scientists, mechanical and chemical engineers, transportation experts and economists or marketing professionals. But more importantly, the greatest challenges we face, including food production, energy, access to clean water, sustainable transportation and so on require the collaboration of diverse disciplines. US universities suffer from barriers and thresholds, which tend to isolate researchers within their own disciplinary 'silos'.

'When you're a hammer, everything looks like a nail' – when viewed from the perspective of a particular discipline, the solutions to problems tend to reflect the attitudes, methods and precedents within that discipline. To be successful, solutions and the organizations which develop them should be an accurate reflection of the challenges, in the sense that the knowledge and expertise demanded by the problem be also represented in the solution; while one of academia's roles in society should certainly be the preparation of a new generation of graduates, and groundbreaking research can certainly be added to this, academia should also be able to address a society's big challenges, not just tiny, esoteric pieces of them. Students visiting German universities learn that multi-disciplinary centers in academia can be effective sources of new ideas and approaches, technologies and insights.

The Fraunhofer-Gesellschaft is one example of the collaboration between industry and university students learn about during their research stay. The Fraunhofer Institutes are an application-oriented research organization geared toward providing benefit to private and public sectors, such as industry, service and public administration. With more than 60 Fraunhofer institutes all over Germany, it is the largest of its kind in Europe. It is financed through contracts with industry, from publicly financed research projects and subsidized by the federal as well as the individual Länder governments in the form of institutional funding. Students in the GI 21 program visit the Fraunhofer Institute in Dresden, which specializes in materials science and technology, while those in the GTP Program in Stuttgart visit the Fraunhofer Institute for Production Automation.

Cultural values play a central role in determining which problems get solved and how

Most of us are well aware that going abroad to work or study is quite different than traveling as a tourist; the tourist takes pleasure in the very things that make one place different from another, finds them oddly curious, amusing or perhaps ridiculous. On the other hand, the student has good reasons to invest extra effort to understand these differences and to deal with them on some level. The student is soon struck by the fact that while societies face largely the same challenges, the actual solutions are rather different. For example, a US student observes that the fences which separate visitors from the animals at the zoo in Germany are typically much less substantial than those designed for use in American zoos. The American speculates, rightly I would say, that American society would hold the zoo operator responsible if an accident were to occur. Germans, on the other hand, have a quite different view of personal responsibility and the consequences of this cultural difference manifest themselves in a wide variety of forms.

During a recent visit to the American embassy in Berlin, a participant in the Global Ingenuity 21 program asked Ambassador Philip D. Murphy what kept him up most at night. He replied that safety and security issues were the most challenging to deal with, owing to differences in Germans' and Americans' views on personal rights versus public safety and welfare. Such stark and surprising differences arise in virtually every problem-solving situation, whether energy, food production, materials use and recycling, urban planning, transportation, economic recession, etc. Students need opportunities to directly experience problem-solving (design) in a cross-cultural setting to develop awareness of the importance of these differences and the skills required to recognize and deal with them. The engineering design portfolio, for which each student develops her own research focus, affords the opportunity to explore and reflect upon culture-specific design solutions. One GI 21 student for example, used photography to examine and document the pedestrian accessibility of places visited and compared these results to the United States.

Engineers solve open-ended problems, i.e. those which have no one, right answer, by first identifying the 'real' problem. The 'real' problem is whatever the stakeholders associated with the problem agree that it is. For example, in the case of nuclear power, Americans may view the problem as how to ensure the safety and quality of life for those living in proximity to the plant. Germans, on the other hand, may view the real problem as how to keep truly dangerous technologies and power out of the hands of government. Thus, the solutions are quite different and arise due to differences in cultural values, traditions and beliefs.[4] Once the real problem has been identified, a number of design requirements are developed. These are also culturally mediated. The student might observe that Mercedes Benz cars don't have pin stripes or cup holders in Germany, whereas most Americans wouldn't buy them if they didn't. Americans view cars as more than a vehicle; it's a place in which you might want to hang out for awhile, sit and enjoy a nice view, listen to the radio, have a conversation, etc. Germans view cars as instruments for getting from point A to point B as efficiently and effectively as possible. Both want comfort, luxury, safety, aesthetic appeal, efficiency, cutting edge technology, performance, etc., but not in the same order.

It is therefore not surprising that we arrive at different solutions to the same challenges. The surprise is that limitations on technological capacity, educational system and level, infrastructure and wealth may have much less to do with the differences than our cultural values.

Excellence in transitioning scientific discovery into technology requires not only engineers, but also skilled craftsmen

An oft-heard lament is that the United States has lost its manufacturing base, usually ascribed to outsourcing and off-shoring. In fact, the loss of manufacturing capability goes much deeper than the economic advantages of cheaper labor and overhead. The condescending cultural view in the United States of manual labor, and arguably, even craftsmanship,[5] has slowly drained such career paths of students. The length of time necessary to acquire the needed skills and the poor compensation incentivize those who might have considered such paths to look elsewhere. Germany provides a somewhat different experience, where respect for skilled craftsmen, such as tool and die makers, machinists, automobile production workers, masons and woodworkers, is much more pronounced. Students in the Global Ingenuity Program learn about this, for example, when they visit the *Gläserne Manufaktur* in Dresden, the transparent manufactory, where Volkswagen assembles the Phaeton.[6] The term 'manufactory', as opposed to the designation, 'car factory', refers to the tradition of German craftsmanship, which is connected to other traditional German values VW represents, such as the striving for perfection, individuality and attention to detail.[7]

Germany provides more educational programs designed to prepare graduates for skilled trades, often two- and three-year programs, than does the United States. As an example, VWGoA, which began production of the Passat in Chattanooga, Tennessee, in May 2011, is working with community colleges there to develop and offer technical degrees. The approach for the Volkswagen Training Academy in Chattanooga consists of standardized technical training found in Germany and throughout the Volkswagen Group worldwide. Chattanooga State has also partnered with Volkswagen to develop a customized curriculum.[8] A partnership with the University of Memphis supports an initiative to recruit, retain and prepare the next generation of specialists in science, technology, engineering and math (STEM). This program links schools with local businesses and reaches down to the kindergarten level to encourage interest in STEM subject areas. VW considers these investments worthwhile and fills a void in the local education infrastructure. Students studying in Germany as part of the GI 21 program see the results of this higher emphasis and respect for skilled workers in the generally high quality of the country's infrastructure, the sophistication and efficiency of manufacturing facilities, the prevalence of in-house machining and prototyping capability, and when discussing the education system and professional opportunities.

Cross-cultural team work, project management and creative problem solving are critical 21st century skills

In response to the question what new skills they had acquired beyond specific engineering skills, many GTP and GI 21 participants said in their evaluations that they improved their ability to engage in creative problem-solving, that they gained greater comfort and confidence in dealing with the open-ended nature of design challenges, and that they noticed a more positive attitude toward risk-taking. In addition, GI 21 participants said that their ability to work in teams was enhanced. Even though the complex dynamics of the problem-solving process seemed chaotic, in retrospect, the students realized that by transitioning back and forth between working with the entire group and then in smaller sub-groups, they were balancing the need to build and maintain consensus with the need to make a large number of decisions, at various levels, efficiently. Some of them related these experiences to enhanced professional and project management skills; an improved ability to manage a complex project, budget resources and deal with a corporate client.

Others emphasized the significance of cross-cultural team work. One student stated, 'The diversity of experiences, perspectives and problem-solving approaches encountered when working across cultures stimulates open-mindedness and a reservation of judgment, which ultimately supports the creative process of engineering design.' Another mentioned a greater understanding for the value of diversity and ability to leverage this diversity as an asset during engineering concept development.

The experience of working in a cross-cultural team raises questions of what it means to be a global engineer – an experience unlikely to happen in a conventional course setting. Students said that working with a corporate client in a different cultural context, helped them develop a more informed perspective on global corporations. One of the most valuable insights was the realization of the extent to which product design can be culture-specific, that the best solution to a given problem depends on where the solution is to be implemented. This raises fundamental questions about cultural values, what we understand as quality of life, and how to integrate them into the engineering design process.

Students experience a 'culture of sustainability'

Sustainable engineering and design are not the domain of any one engineering discipline or even a discipline in its own right, but must become integral to the thinking and approach every engineer uses to solve problems in the coming decades. Students studying or working in Germany, even for short periods, experience a pervasive and significant awareness and level of responsibility among Germans to exercise sustainable thought and practice, that is, they become aware of the 'culture of sustainability' that exists in Germany. The same can be said for a number of other study abroad destinations, popular among US students, including France, Spain and Japan. In a culture of sustainability, children grow up exposed to attitudes, awareness and actions implying the acceptance of certain truths, such as the Earth's natural resources are finite, and that destruction of habitat, species, topsoil, etc. and global warming are the result of human choices and behavior. The significance of these problems and the enormity of the challenges in dealing with them are treated pragmatically, as reflected in educational programs and curricula, government-funded studies and policy decisions, as well as the choices ordinary citizens make in their everyday lives. It is bewildering for many Americans to experience this, for whom the very nature of these things, as fact or ideology, is still subject for debate. Whether or not Americans studying abroad in places like Germany come to accept sustainability as a serious question, and one requiring smart, prompt action, the key is that they are confronted with an entire society which has accepted it and are thus forced to question their own assumptions and attitudes about it. In the conclusion to her engineering design portfolio on 'Nachhaltigkeit' (sustainability), in which she discusses different examples ranging from the bottle deposit to the ecological design of the Reichstag, one student reflects on her role as an individual and as a global engineer: 'By changing my own view on sustainability, I can have more of an influence on other's views and practices. Traveling abroad was a great way to expand my understanding on sustainability and see how my own country and community may be lacking'.

The organization of educational systems and even their approach to learning are culturally mediated

During their stay, students learn about the German education and university system, and Germany's approach to research, development and collaborative problem solving, which inevitably leads to comparisons with their own academic background and culture. These insights are especially valuable since they provide new perspectives on fundamental questions such as what it means to be an engineer, what sort of problems do engineers work on, what is the aim of engineering and how to think of engineering as a profession.

Engineering schools in the United States are comprised of departments, each department conceived around a body of (primarily theoretical) knowledge. The faculty (consisting of assistant, associate and full professors) have responsibility for meeting the core teaching requirements of the department, as well as for their own specialized courses and research areas. There are a greater number of professors in the US approach, but each has a smaller number of coworkers to supervise and a smaller total research budget. Individual professors are less likely to work in a multi-disciplinary way, tending to focus more on their narrow area of specialized interest.

Through conversations with German professors and students while visiting the host university and various research centers, American students discover that German engineers are more specialized upon graduation than in the United States, that an engineer in Germany is less likely to think of herself as a 'mechanical engineer' or 'electrical engineer', but as a 'building construction analyst', 'energy economics engineer' or 'power plant design engineer'. The university/engineering school in Germany is organized around research themes, such as 'composite materials', 'semiconductor technology' or 'aircraft propulsion'. As an example of this degree of specialization, the students were invited to attend a 'sustainability lecture'; the lecture turned out to be on efficient compressed air flow and handling in a factory. As one Global Ingenuity 21 student put it, 'the German system appears to be problem-driven while the American system is theoretical, followed by the corporate search for applications and economic value'.

A single professor in Germany, at the top of the pyramid, oversees all activities of his/her 'institute'. Departments, referred to as 'faculty', are composed of a number of such research institutes. Due to the high degree of individual specialization, the research institutes are multi-disciplinary. As an example, a partner with the Global Ingenuity Program, the Lower Saxony Center for Automotive Research (*Niedersächsische Forschungszentrum für Fahrzeugtechnik*, NFF) is an interdisciplinary research center at the TU Braunschweig.[9] The interdisciplinary research team, consisting of 15 member organizations, with professors from TU Braunschweig and other German universities, is aimed at the development of vehicle-oriented technologies.

Engineering is a cultural activity and language is the gateway to cultural understanding and engagement

The experience of learning language helps sensitize students to cultural differences prior to their study abroad and makes them more aware of their own language. Basic German language instruction begins six weeks before the GTP and GI 21 programs' departure. As far as language learning itself is concerned, the goal is to familiarize the students with the German language. They meet on a voluntary basis, twice a week for one hour. The focus is on speaking, with only very basic grammar formally introduced and taught. After the instruction, the students are able to say a few basic phrases, such as their name, where they are from, order at a restaurant and ask directions.

English is the main language during the GTP and GI 21 programs: all lectures and discussions are conducted in English. The bilingual ability of the German students helps with translation when necessary, and brings the American and German students closer as a group. In their evaluations, the UVA students remarked how the mutual awareness of the challenges and effort of learning one another's language, culture and traditions leads to respect. This is a basis for trust and is therefore foundational for the collaboration needed to solve global challenges.

Basic language training and study abroad raises a new awareness of their own language and their rhetorical approach to expression. Language is no longer taken for granted, as a mere instrument to convey meaning, but is frequently questioned due to linguistic and cultural barriers. The US students have to adapt to their non-native speaker audience, which means on the most basic level that they have to speak more slowly, that they refrain from using idiom and colloquialisms, and phrasal verbs etc. Students are challenged to step back, explain and clarify what they said and to consider things from various perspectives. This prolonged intercultural dialogue helps them define more precisely what they are thinking and planning, and how to proceed. Language and cultural barriers, while leading potentially to confusion and inefficiency, may work to participants' advantage in the engineering design process.

Especially with regard to problem statements demanding more creativity to solve, language can play a crucial role in the engineering design process. For example, to be able to more narrowly define the safe use of the Internet while driving, the students worked with metaphors to describe their expectations of the car and the interface, in terms such as 'Fahrvergnügen' or 'best friend' and Internet sites as 'fish' caught while net-fishing. This approach not only helped the design team solve the problem, but increased their awareness and appreciation that language is symbolic of deeper, cultural values. In the final evaluation, some students said that they developed a greater appreciation for the value of metaphor, not only during the creative brainstorming phase, but as a means of explaining complex ideas during presentation.

Summary

While one might assume engineering to be a 'universal', objective language, much like mathematics is purported to be, in fact, the education and professional practice of the engineer and the process engineers use to solve open-ended problems are profoundly affected by the cultural values, traditions and beliefs which define a society. The skills needed to successfully transcend these differences are learned through practice and direct experience of engineering within culturally diverse settings. Though quite distinct in organization and approach, the two short, intensive engineering study abroad programs presented provide significant benefits for the professional development of the global engineer.

Notes

1. See, for instance, the 'Criteria for Accrediting Programs in Engineering in the United States', http://www.ele.uri.edu/faculty/daly/criteria.2000.html.
2. G. Tryggvason and D. Apelian (2006) 'Re-engineering engineering education for the challenges of the 21st century', *JOM,* 58(10), 14–17.
3. Wm A. Wulf, and George M.C. Fisher (2002) 'A makeover for engineering education: Today's engineering schools are not preparing their graduates as they might for useful practice in the 21st century', *Issues in Science and Technology,* (National Academy Press), 18(3), 35–39.
4. Gary L. Downey et al. (2006) 'The globally competent engineer: Working effectively with people who define problems differently', *Journal of Engineering Education,* 95(2), 1–17.
5. Matthew B. Crawford (2009) *Shop Class as Soulcraft: An Inquiry into the Value of Work* (Penguin Press, New York).
6. Website of *Die Gläserne Manufaktur,* http://www.glaesernemanufaktur.de/.
7. Website of *Die Gläserne Manufaktur,* http://www.glaesernemanufaktur.de/de/idee/manufakturbegriff.
8. Website of Volkswagen Group of America, http://www.volkswagengroupamerica.com/chattanooga/partners_in_education.html.
9. Website of Technische Universität Braunschweig (Niedersächsisches Forschungszentrum Fahrzeugtechnik, NFF), https://www.tu-braunschweig.de/forschung/zentren/nff.

11
Designing for the Anthropocene: The Duisburg Rhine Park

Jörg Sieweke

How can we construct a socially and ecologically sound world which we are increasingly responsible for shaping ourselves? How can the profession of landscape architecture help to mediate the process of modernization and urbanization? This chapter elaborates on a proposal for the adaptive reuse of a brownfield site. The strategy discussed here suggests a holistic design approach to deal with contaminated post-industrial sites in a productive and pragmatic way. Instead of neglecting and covering the industrial remains, the management of the industrial heritage and contaminated ground is understood as a challenging design opportunity. The clean-up and redesign of the site is discussed in the context of a critical revision of the meaning and identity, both past and present, of post-industrial landscapes.

The advent of the steam engine marks the beginning of the *Anthropocene*, an era in which the impact of human interventions on the environment has reached a global scale. The term 'anthropocene' captures this unprecedented scale of environmental transformations by humankind. Formerly only 'eons', or geologic ages of the earth, captured this scale of transformation. The pride of the heroic pose to successfully control and subdue nature is being followed by a shudder, facing the unintended, but self-induced, consequences of climate change and other environmental problems.

Countries that like to think of themselves as developed have begun to critically assess the side effects and shortcomings of the project of industrialization and modernization, which originally promised freedom and liberation to humankind. While the consequences of global warming, such as rising sea levels are ubiquitous, they remain fairly abstract since they rarely affect us directly. From the perspective of modern industrialized societies, brownfields are more immediate and tangible than the predicted global warming curves and the locally imperceivable temperature rise. In densely populated Western Europe, these post-industrial sites are going through a recycling process to integrate them in future land use patterns for the next generation of value-added production. In the past, commodities such as coal and steel were extracted and processed here. The sites were left behind in a degraded state, as remediation was not provided by the corporations responsible for their degradation. The waste and dross of the process were dumped on the site without further concern for their toxicity, and the costs for the clean-up have been passed on to the general public. The logic is not much different from that of the financial industry crash of 2008: gains are privatized while the responsibility for toxic assets is passed on to society at large – a burden for taxpayers of the current and future generations to deal with.

Only recently has the 'polluter pays principle' (*Verursacherprinzip*) legislation significantly changed the legal framework toward an internalization of these previously externalized costs. This new principle of liability helps to prevent further cases of abandoned brownfield sites, but does not address the larger past incidents which are 'grandfathered in' and remain the problem of the public. Lars Lerup refers to 'stim and dross' as two alternative outcomes, where islands of 'stim' receive all the attention, funding, and investment and the surrounding environment of 'dross' is characterized by disinvestment and neglect. Allen Berger builds on this concept with the publication of his 2006 book, *Drosscape*, which refers to vast areas of contaminated, post-industrial brownfield in the United States and elsewhere. This chapter offers a strategy to recognize and signify the heritage of the industrial era and to make its remediation and the limits of remediation perceivable to a larger audience. It embraces the Lerup concept of 'dross' and challenges the concept of landscapes of neglect.

The strategy discussed here was originally developed in the context of an international design competition entry. The design, conceived in 2005 for a site in Duisburg located on the Rhine River, integrates the engineering aspects of a remediation project with the conceptual design strategy for a future urban landscape based on an assessment of the potential to negotiate the contested ground. Remediation is typically understood as either a technical or biological process of containing, securing or retracting contaminants from the ground in order to prevent them from coming into contact with air, water or users of the site. Depending on the sensitivity of the designated future use, benchmarks for certain toxic substances are legally binding and must be met.

An underlying assumption is often that as long as toxic substances are immobilized, they will not be harmful.

IBA Emscherpark

The International Building Exhibition (*Internationale Bauausstellung*, IBA) is a quasi-governmental regional consortium that acts as an agent for change and innovation in regions challenged by economic transformation in Germany. Many projects implemented by the IBA Emscherpark, such as Duisburg-Nord or Zeche Zollverein, are recognized internationally as best practices for post-industrial regions. The IBA typically focuses on a particular region for 10 years to develop goals and an agenda to steer the process of structural transformation. IBA projects must qualify by meeting the consortium's standards of innovation and quality to receive public funding. The consortium prioritizes parks and public greenway projects to establish a landscape armature, enabling a new sense of belonging and connectivity. Public space is the spine that attracts and ties in investments of the private sector, allowing value to be realized in the 'recycling' of brownfield sites, rather than moving out into unspecified areas of suburbia or greenfield development.

The Duisburg Rhine Park project site is located in the Ruhr area, an urbanized and industrialized region in the western part of the country. It represents a typical site, characteristic of a larger regional process of structural change comparable to the Rust Belt or the 'German Coast' along the lower Mississippi. The economic restructuring (*Strukturwandel*) through the 1990s in the Ruhr region in Germany is recognized as a successful example of post-Fordist transformation. Companies such as Mannesmann successfully transitioned from heavy industries of coal and steel to communications and information technology and services while staying in the region. After 20 years of ambitious redevelopment, the Ruhr has managed to overcome the stigma of a deteriorating heavy-industry region. A successful transition to a third sector of service, knowledge, and information industries has begun in a reinterpreted, post-industrial setting.

Despite the largely successful economic transformation in the Ruhr area, the post-industrial landscape shows many physical and territorial traces of its previous identity. Some traces may be considered cultural artifacts and others, toxic dumps. Typically, these opposing characteristics are inseparable. The IBA Emscherpark's strategy in this transformation process is based on the creation of a framework of public open spaces and greenways alongside preserved infrastructure, such as canals or rail corridors. In doing so, it creates places and builds vantage points marking qualitative successes in renewal within the industrial landscape.

One of the lessons of the IBA Emscherpark concerns cultural values. It signifies a paradigm shift from the notion of moving beyond the industrial age by neglecting and eradicating its historical traces, to a strategy that sees value in this particular history and its industrial heritage. The IBA has helped to attribute value to the industrial identity of the region by tying many cultural activities and arts programs to redevelopment project sites. The interpretation of the sites and built structures is recoded in a positive sense, which then helps to attract new commercial uses. Cultural identity is considered a soft parameter for locating businesses, comparable to proximity to golf courses in other locations. Contemporary park design helps to create these characteristic associations.

One popular precedent of the early IBA projects in the 1990s is another park close by, the Duisburg-Nord Landschafts Park. It incorporates many authentic elements of its place and even the material structure of the steel plant towers themselves. Designed by landscape architect Peter Latz, the Duisburg-Nord park design became iconic for its contextual design approach. It stood out among other competition entries at the time by working with the given structures of the site, incorporating and reinterpreting existing topography as a foundation for the park. Most other design proposals utilized the conventional 'tabula rasa' approach, imposing new geometries and new structures over a site mistakenly understood as a 'clean slate'. Latz's approach was different; the reinterpretation of coal bunkers into enclosed garden spaces worked well, with minimal insertions of pathways and access points. A natural gas tank structure was repurposed as a scuba diving tank and the alpine hiking club appropriated a section of steel plant for climbing practice facility.

Duisburg Rhine Park

Like the Duisburg-Nord park, the brownfield site considered here should be developed as a park in order to build a new 'address' or sense of place, for a new mixed-use urban development along the riverfront. The riverfront was not publically accessible during the past 150 years of the industrial era. The opportunity to reconnect the urban fabric back to the Rhine waterfront led the municipality to hold an international landscape architecture and urban design competition. The competition brief states that a mixed-use district, including a riverfront park, should mark this new address with public space. One core question of the competition was about the character of the public park that would best suit the site. The brief asked entrants to design a new top layer as a 'green blanket', assuming the contaminated layers would be capped and fully remediated already. However a further challenge identified when developing the Duisburg Rhine Park scheme became how to reference the industrial history and heritage of this particular site in an archeological sense, acknowledging that all traces would be buried underground and no structure would remain above the ground.

The history of the site and its relationship to the Rhine is quite dramatic. The city of Duisburg is located at the confluence of the Ruhr and Rhine rivers. In the 12th century, when the Rhine dramatically changed its course after a large flood event, new territory was suddenly created on the high steep bank of the river, between the city and the river's new path. This development was followed, beginning in the 1830s, by a period of heavy industrialization. The site has seen numerous industrial uses, including a coal mine, a coking plant, a steel plant and multiple harbor basins. This particular 150-year sequence of destruction and rebuilding of heavy industry has been materialized in an industrial archaeology of layered debris, foundation and utilities, all remaining below ground. Erected on the rubble and debris of two world wars, the current steel wire rolling facility owned and operated by the Indian company ISPAT will eventually be relocated to free the entire area from its industrial past – at least above ground. Invisible under the surface, the material traces of various foundations and contamination, up to 12 meters in depth, will remain. The stratified anthropogenic topography of this vast industrial dump masks the natural, terraced geology of the Rhine Valley.

Critical reading of the context

Design strategies often emerge from a critical assessment of the existing context, as well as the conventions formulated in the competition brief itself. This entails raising larger questions that may reach beyond the limits of the site and the description of the problem posed. In this respect it should be noted that Duisburg, like most cities in the Ruhr region, is not lacking green space, in quantitative terms. Quite the contrary, the city offers an abundance of nondescript, generic green space. Further, the post-industrial region itself has vast areas of fallow brownfields with spontaneous vegetation, often referred to as 'post-industrial nature' or 'industrial forest'. Both qualities are ubiquitous in the region and are not intended for the Rhine Park design scheme. Rather, our proposal raises the question of what kind of green might be most appropriate here.

Should the design affirm the idea of 'discrete green' being a compensatory requirement for unsatisfying and unhealthy living and working conditions? The conventional concept of the classic modernist urban planning doctrine, represented by Le Corbusier's Athens Charter, spatially separates the areas of daily life into work, residence and recreation. They are considered conflicting uses of space and therefore spatially divided into discrete mono-functional districts. In this understanding, 'green' represents one land use category of its own, isolated in a programmatic box of recreational 'zones'. Our proposal employs a more diverse concept of various constructed natures, synthesized into the concept of a new, hybrid type of park. The recovery of the landscape envisioned here is based on a more thorough rehabilitation of the site with its ecological functions on one hand, and an obviously artificial surface for the on-site, controlled and capped area of the landfill on the other hand.

Background contamination and the blanket of neglect

How do we manage the inherited contamination in the context of the scale and extent of an entire industrialized region? In light of the vast scale of brownfields and the lack of funding for cleanup, pragmatic and opportunistic strategies are becoming the default. One inconvenient truth confronted during the course of this design was the fact that the contamination affecting the site could not be extracted without considerable economic and technical resources. In other words, the process of large-scale contamination is not practically reversible. This assessment poses a tough challenge for a culture that is still deeply rooted in the modernist paradigm of control and order, with a strong belief in technical 'can-do' (*Machbarkeit*).

Because contamination can only be relocated, but not resolved, the conventional response is to cap the site. The prevailing remediation procedure may be criticized as the 'green blanket of silence' strategy. Large areas of contaminated soil are simply covered with impervious layers of PVC sheets. A three-foot layer of new topsoil, referred to as the 'cap', is then used to cover the site. This practice necessitates the management and monitoring of the surface itself and of all of stormwater above it. New uses are introduced on top of this blank slate. These new uses are typically exposed to higher risk than they would be in other locations due to the possibility of being exposed to contaminants below. The default conditions laid out in the brief consider the engineered remediation project prior to and separate from the final landscape design project, with each having a separate budget.

Proposed recycling of the territory

Our proposal for the site avoids splitting the treatment into a technical remediation first and design of a green blanket second. Instead, the proposal suggests a holistic, integrated approach which implements the ecological recovery in the design and draws from its sectional properties as an intrinsic compositional component. Our study of the site determined the cut and fill management of the earth and how it could be implemented in the site remediation concept. This analysis allows a mapping of the unaltered geologic layers as well as the anthropogenically altered geological layers. The unaltered layers are of Quaternary or Tertiary deposit, formed by glaciers and later by rivers carrying various grain-sized sediments, whereas the anthropogenic strata are precisely the industrial archaeology of derelict fill and dross. The geologic survey sections indicate a variation in the depth of the contaminated anthropogenic layers, ranging from 1 up to 12 meters, or more, in depth.

A pragmatic and feasible approach would be to limit the interventions to a depth of two meters. Up to this depth it is reasonable to cut out all contaminated materials if they reach the intact quaternary river gravel (layer –1). The excavated contaminated soil will then be stored and secured on top of the already more severely and deeply contaminated areas that would need to be capped anyway (layer +1). The structure of the park is therefore determined by three distinctly different layers that reflect the recovery of the site as well as future uses, represented by different cultural interpretations of nature (Figure. 11.1). The new topography is derived from three autonomous layers with varying elevation and surface material cover that then define the qualities of the park:

The +/– 0 layer is 100 per centgeneric. This layer serves as a reference of 'generic green' found in most parks and gardens in the city. This represents the conventional blanket approach of putting a layer of fertile earth atop an impervious PVC seal of the contaminants below. It is the default remediation suggested by the brief and provides a default elevation to understand the distinctly different, sectional logic of the following two layers.

The +1 layer is 100 per cent artificial. This is the extruded and capped landfill, designed as a purely artificial surface. The impervious cap surface is like a plaited cloth. Within the striated folds 100 promenades provide linear spaces for various forms of active or passive recreation. This landscape does not neglect its artificial nature; rather, it reveals the

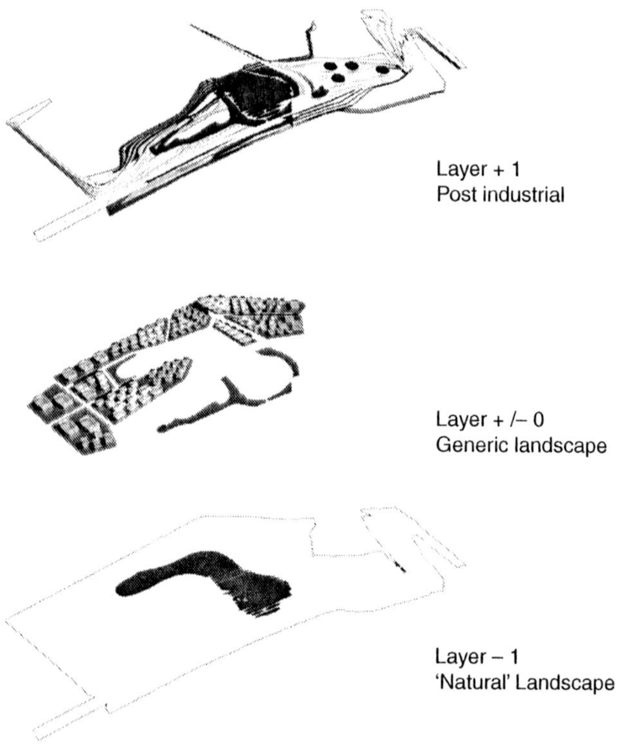

Figure 11.1 Design concept: Three landscape layers

intrinsic character of its artificial construction and the limited depth of its surface. This layer of the park is intended to be read as an artifact, including its engineered stormwater management.

The –1 layer is 100 per cent natural. It reestablishes nature by excavating the quaternary gravel layer, giving it back to the landscape in its aesthetic and ecosystem functions. This includes the infiltration and recharging of the groundwater and aquifer. Analogously with the opposite riverbank, it is an inhabitable open space along the river in the form of a wide open meadow that allows for a wide range of activities. In this way, a 50-acre area can be recovered by removing the top two meters of contaminated soil. In addition to improved filtration of stormwater runoff from the recovered layer of river, the benefits of this solution include cost savings from reducing the capped surface area.

Generations of green

The three layers addressed in the proposed scheme represent three distinctively different characters of *green*. The conventional 'green blanket' (layer 0) that typically resembles a high-maintenance generic lawn with playing fields on top of a cap is cited and critiqued in order to contrast the particularities of the proposed layers +1 (hyper-nature) and –1 (true nature) against the default condition. The layers reflect the critical position that the city of Duisburg is not lacking generic green space; rather, it is lacking in much more important, identifiable and characteristic open space as a 'cultural landscape', signifying and referencing its industrial past as a vital reminder of it.

Ethical standards and social equity

The proposed park establishes a variety of specific situations with distinct qualities and habitats that may provide numerous options for a broad ethnic mix of residents (Figure 11.2). The goal is to avoid the temptation to provide one generic setting for all users and their demands. Duisburg's population consists of multiple generations of immigrants from over a hundred different countries. During the period of industrial growth, large numbers of 'guest workers', typically from southern Europe, were incentivized to move to Germany to be employed in the industrial job market. The ethnic mix soon became a cultural characteristic of the Ruhr region, bound together by the hard labor ethic, self-esteem and other elements of a unique culture. Soccer became an important element in helping integration. The new park design features an open meadow to serve as a vast public ground that can be used for games. It can also be used for barbecuing, a popular and common outdoor activity for the Turkish community in particular. The more secluded areas provide intimacy for uses geared toward individuals or smaller groups. The linear promenades are partially paved to allow for various wheeled sports, such as rollerblading, skateboarding and BMX biking. The concave folds in the plaited surface will harvest moisture and dust. Depending on aggregation and soil and seed accretion patterns, successive plant communities are anticipated to emerge spontaneously, creating yet another way to experience the park.

Phyto-remediation

The design can also be understood as a critique of the trend known as 'green-washing', where random superficial and unsustainable practices

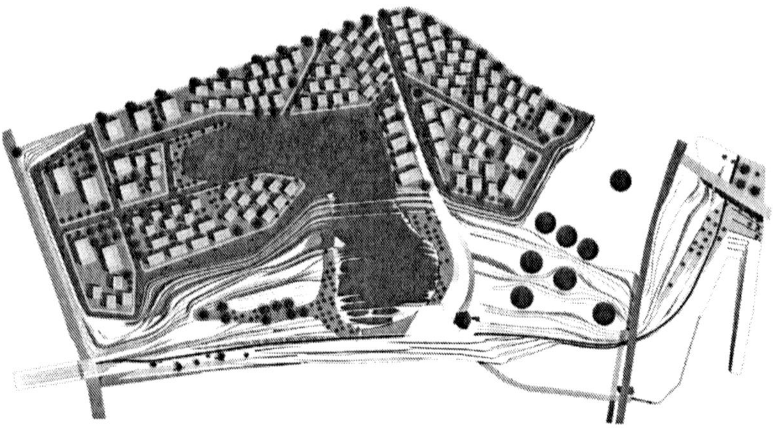

DUISBURG RHINE PARK:
Recycling the territory:
The concept is to avoid splitting the treatment into a technical refitting and a design work. In contrast the proposal introduces a holistic approach that implements the ecologial recovering in the design and makes an immanent qualitative difference.

Implemented sanitation concept:
Based on the analysis of the given geogenic and anthropogenic conditions a mapping of their polentials is made.
The geological sections show a variation in thickness of the contaminated soil layer.

It varies from 1–3 m up to 12 m, up to a depth of 2 m all contaminates are removed and the natural gravel comes to the surface again.
The contaminated soil will be stored and secured on top of the lready heavily loaded areas.

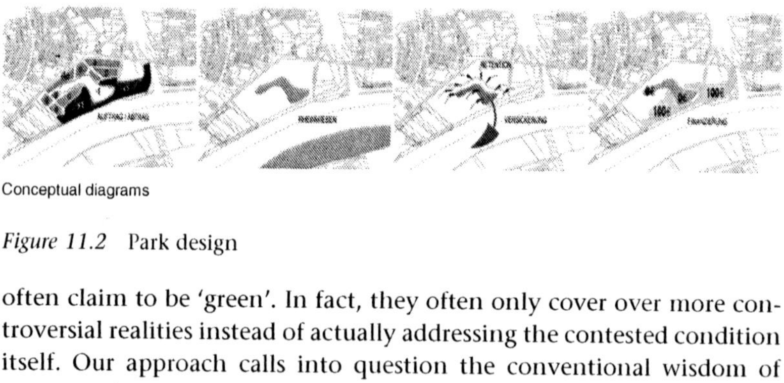

Conceptual diagrams

Figure 11.2 Park design

often claim to be 'green'. In fact, they often only cover over more controversial realities instead of actually addressing the contested condition itself. Our approach calls into question the conventional wisdom of green as the smallest common denominator for harmony. Instead of providing the reassurance of an allegedly unproblematic green lawn, the project engages the full sectional spectrum of a deeply problematic site and makes it perceivable as a fact. It therefore provides a more thorough resurrection of the territory and its meaning.

The proposal manages all material remediation processes in situ, or within the site itself. Phyto-remediation is a relatively cost-effective procedure to break down toxic substances in the soil. It provides an alternative to high-tech procedures like soil-washing or thermal treatment, which are faster, but much more expensive. Organic substances such as oil derivatives are broken down by bacteria which are stimulated by the specific soil treatment. The soil is cultivated to achieve specific moisture and oxygen milieus, pH-levels and other parameters that positively affect the bacteria's productivity. In addition, so-called land farming practices such as aeration and the turning and irrigation of the soil catalyze the breakdown of toxic organic matter by stimulating certain bacteria. Other strategies involve using plant species that take up contaminants with the water and store them in their tissue. In particular, fern species have been identified as highly productive in this regard. The fern, together with the contaminants, can be harvested after removal of the toxic heavy metals from the soil. This and other extraction or bioremediation procedures may be tested on the site as a publically accessible demonstration of onsite bioremediation practices.

Buttes chaumont

The strategy to engage the contested history of a site as part of its future identity has historic precedence. The renowned Buttes Chaumont Park in Paris can be interpreted as the world's first post-industrial park. Commissioned in 1867 by Napoleon III for the Exposition Universelle, or World's Fair, it was built on the site of a former quarry that provided the stone to help build Paris. The dramatic topography of the quarry site was not leveled, but rather exaggerated to create the spectacular experiences of a grotto, a steep ridgeline, a waterfall and an abyss as compositional components of the park's design. The then relatively recent innovation of a railroad line also crossed the park before heading into a tunnel. This design strategy embraced the early industrial presence and developed its character further as integral elements. The steep climb and the railway tunnel mouth can still be experienced today. Buttes Chaumont can be understood as the first post-industrial park that integrates the presence of the site and the public infrastructure of its time. Reflecting on contemporary parks, the concept of Buttes Chaumont stands out as an early example of what parks should strive to achieve today. It does away with the concept of harmony and diminishing contrasts or contradictions, embracing them and bringing them to the forefront of the design. Buttes Chaumont has proven itself, and remains a very popular and actively used park of contemporary Paris. Today, with its design unchanged, it remains modern in spirit.

Public parks serve an important socio-cultural purpose in every free and open democratic society. Parks are the stage on which one can playfully and without any specific purpose meet and engage with the unknown 'other', helping to constitute the understanding of one's own role in society by mirroring it in the encounter of the other. The collective sum of these individual encounters constitutes a liberal culture in real-time.

Undisciplined design

The historic example of Buttes Chaumont and the recent Duisburg cases discussed here demonstrate the capacity of design practice to function as field of innovation which develops and reflects an understanding of the state of contemporary culture. At its most innovative, design thinking is *undisciplined* in two important ways.

The first way is that design thinking is holistic, meaning that is does not belong to any one specific profession or specialty. Rather, it is one of the few remaining generalist problem-solving approaches. In broadening the conversation and scope of a project to other disciplinary fields, including economic, ecological, social and cultural perspectives, the designer and his or her design embrace both the historic and contemporary cultural implications of the design project.The crucial moment in the design process is when the designer must synthesize and condense all the relevant information with the aim to bridge the many fields and pieces of information into one coherent scheme.

The second way that design is undisciplined is its non-conformist attitude, or critical thinking. It questions conventions and establishes new modes of thinking. The critical designer identifies unquestioned assumptions hidden in the task and allows him or herself to experiment and investigate alternate routes. These playful explorations of alternative paths are driven by the hypothetical lifting or even breaking of long-standing conventions and rules. In some cases, the implementation and testing of alternative design schemes have proven to contribute to the emergence of new rules, regulatory insights and continued innovation. The design proposed for the Duisburg Rhine Park harnesses the undisciplined nature of the design process as a path to insight and knowledge.

12

Eco-Revelatory Design: An Approach You Can Bank On

Eugene Ryang

In 1998, *Landscape Journal* published a special issue entitled *Eco-Revelatory Design: Nature Constructed/Nature Revealed*. This issue showcased conceptual projects in the fields of landscape architecture, planning and land management projects that shared the common objective of using artistic design interventions to bring transparency to a site's natural and/or constructed systems. Embracing a multidisciplinary approach, the exhibits 'revealed and interpreted ecological (and engineering) phenomena, processes and relationships' through symbolic expressions and aesthetic applications to the site (Helphand and Melnick, 1998, p. x).

The overarching goal of these conceptual projects was to make the typically invisible functions and processes of a site visible through thoughtful and beautiful design applications and interventions, hence the term 'eco-revelatory design'. The underlying assumption is that once these functions become transparent, visitors to a site become more conscious and thus more sensitized to the everyday landscape. According to this logic, if participants can visualize and experience environmental processes and phenomena, then they are better able 'to appreciate, evaluate and make wise decisions concerning them' (Helphand and Melnick, 1998, p. x).

Too often, the functioning or dynamic component of a site – either as an existing ecological system or as a human-made engineered environment – is hidden from visitors/participants. As a result, visitors/participants can become disconnected from the bio-physical aspects of the world that surround them. Once disconnected, neglect and ignorance of ecological and environmental concerns usually follows. As Sym Van Der Ryn and Stuart Cowan maintain in their book *Ecological Design*:

> We have culverted the creeks, paved the wetlands and built on the farms, orchards and meadows that once nourished young minds.

> We have rendered both nature and the consequences of our own technologies increasingly invisible ... Many of us live in cities where both ecological and technological processes are hidden from our everyday awareness. The designed environment does not reveal to us how technology supports us and how, in turn, it is interconnected with the natural world ... There is a pernicious cycle at work here. As our system of food, water, energy, waste and sewage have grown ever more intricate and hidden, it has become more difficult to understand or question them. As nature has receded from our daily lives, it has receded from our ethics.' (Van Der Ryn and Cowan 1996, p. 161)

My experience, as principal of an ecological planning and engineering firm, supports Van Der Ryn and Cowan's commentary on the absence of ecological and technological processes in everyday life. Conventional planning and engineering practices often deliberately conceal the inner and outer workings (or infrastructure) of sites. For a multitude of reasons (i.e. maximizing land utilization capacity; prominence of vehicular accessibility; misguided engineering strategies, etc.) standard site plans obscure (usually by burying) much of a site's dynamic modes of operation.

When it comes to site planning and engineering, conventional practice seeks (1) to clear and denude sites of any and all plant and animal habitat in order to develop as much land as possible; (2) to thoroughly flatten the ground plane (remove the contours of the land including water courses and drainage ways) in order to create a tabletop surface for convenient layout and construction of buildings, roads, parking lots, etc.; and (3) as mentioned above, to render the ground plane neutral for ease of burying all natural and/or engineered infrastructure beneath it. What is thought to be the most economic and efficient means of engineering and construction often ironically turns out to be more costly both economically and environmentally. In sum, active engagement with and deference to a site's living processes typically are not parts of the design engineer's equation.

Practitioners of eco-revelatory design believe that making ecological and engineered processes more visible is a way of reacquainting us with the symbiotic relationship between culture and nature. This awareness, they argue, will help inform citizens about the ecological consequences of their actions and activities. Transparency alone, however, is not enough to create this kind of awareness. Eco-revelatory design advances the idea that transparency must be coupled with artistic intervention to 'de-center' status quo perspectives and offer alternative ways of exploring and understanding that which is typically hidden from our everyday lives. As Elizabeth Meyer asserts in her manifesto, *Sustaining Beauty. The Performance of Appearance*:

> It will take more than ecologically regenerative designs for culture to be sustainable, that what is needed are designed landscapes that provoke those who experience them to become more aware of how their actions affect the environment and to care enough to make changes. This involves considering the role of aesthetic environmental experiences, such as beauty, in re-centering human consciousness from an egocentric to a more bio-centric perspective.' (Meyer 2008: 9)

Meyer elaborates on the 'amplification and exaggeration of experience' in artistically exploited landscapes. There is, she asserts, a restorative component to experiencing the natural processes, structures and functions of beautifully constructed sites. She cites Elaine Scurry on the profound impact of the 'performance of beauty'.

> At the moment we see something beautiful, we undergo a radical decentering. (...) It is not that we cease to stand at the center of the world, for we never stood there. It is that we cease to stand even at the center of our own world. We willingly cede ground to the thing that stands before us.' (Scurry in Meyer 2008: 13)

The premise is that eco-revelatory landscapes are not merely ecologically sensitive. They artistically frame functioning systems in order to create experiences. They compel us to give pause and reflection to seemingly ordinary spaces, persuading us to consider, perhaps, the more abstract nature of everyday processes. Meyer makes the call for 'reinserting the aesthetic into discussions of sustainability'. Once inserted and part of the ecological design process, these environments become hybrid landscapes that can lead to 'attentiveness, empathy, love, respect, care, concern and action on the part of those who visit and experience (them)' (Meyer 2008:19).

Along with eco-revelatory design theory, our planning and engineering team has been equally influenced by practitioners with similar philosophies. Such influences include German artist and landscape architect Herbert Dreiseitl and German landscape architect Peter Latz (see Jorg Sieweke's Chapter 11 in this volume). Dreiseitl and his firm have created precedent-setting installations incorporating art and design with sustainable rainwater and stormwater management. Their regenerative 'waterscapes' operate at multiple scales – from water features in small city squares and plazas in Germany, Australia and the United States to large-scale water resource management projects in cities such as Singapore. Another inspiration for our work, Peter Latz, has focused much of his efforts to transform derelict and abandoned sites into parks and recreational facilities. His designs utilize plants for biological uptake and remediation of heavily industrialized soils. Dreiseitl and Latz's work combine form and function in a transparent and aesthetically compelling manner that allows people to connect to the biodynamic operations of the site. Their success has encouraged us to do the same with our own projects.

Our studio is in Charlottesville, Virginia, a historic college town located in the foothills of the Blue Ridge Mountains. Home to the University of Virginia and numerous design, planning and engineering firms, Charlottesville has an abundance of local projects that both exemplify and contradict the ideas expressed by eco-revelatory theorists. This is particularly evident with projects that involve the handling of water on site, i.e. stormwater management. It is understood that stormwater management is one of the more important, if not most important, tasks of civil engineers and ecological planners/designers. Visiting local sites, it becomes quickly apparent which firms seek alternative strategies to reveal and enhance a site's stormwater systems and which ones continue to operate under the conventional pipe-to-pond and pipe-to-stream paradigms. Carol Franklin, principal of Andropogon Associates – a prominent ecological planning and design firm in Philadelphia – maintains that 'creative stormwater management is the key to every site ... stormwater is the tail that wags the dog. It determines site design. [Generally] it's left to engineers, and more engineers are usurping the aesthetic by creating plumbing systems'.[1]

The remainder of this chapter focuses on an unassuming bank project just west of town. I use this example for three reasons. First, it was one of my earliest experiences in adopting an alternative, low-impact approach to stormwater management. That is, our team used an innovative methodology that introduced less harmful nutrient and sediment loads into our waterways, while maintaining the same stormwater quality and quantity requirements of conventional engineered systems. This approach proved to be more ecologically beneficial for our impaired local watershed, which is part of the larger and severely compromised Chesapeake Bay Watershed.[2] Second, I was fortunate enough to work on this project as a principal in the firm McKee Carson. This particular project enabled me and my future business partners to hone our interests in eco-revelatory design and launch our own multidisciplinary firm in Charlottesville (Waterstreet Studio). Finally and most importantly, this project is distinctive in its accessibility as an everyday landscape. Frequently, we find that the more thoughtful design work is part of private commissions – the provenance of privileged clientele – with high profiles and low visitor visibility. This project, by contrast, is quite public, modest in scale and offers the opportunity to illuminate functioning engineering and ecological processes for a broad cross-section of citizens.

Our charge was to take an abandoned lot with a damaged urban stream (Figure 12.1) and to create a building site that would accommodate a bank and its accompanying structural and infrastructural requirements. We could have applied the standard program and designed a conventional site plan with buried infrastructure and a large parking lot. It would have been easy to work out on paper as we had performed similar work dozens of times. This normative strategy was, in fact, one of the initial proposals put forward during the concept phase of design. In the conventional scheme, burying the stream that bisected the site and putting it into a box culvert would have made for expansive parking areas – highly prized infrastructure for the typical commercial project. Fortunately, it did not take much to convince the client that a different design, including an alternative stormwater management approach, could create a more transformative site experience; one that would not only preserve the site's natural asset (the stream) but also properly filter

Figure 12.1 Bank site and Meadow Creek Stream *before* design implementation

and remediate the stormwater pollutants conveyed from the building, parking lots and travel ways. We could accomplish this and still create enough parking required to accommodate customer demand.

Despite a little more creative design and outside-the-box engineering effort during the planning stages, the alternative proposition would cost the same amount to build, be better for the environment and be more attractive for business. So with the blessings of the client, from concept phase through construction, every detail was worked out to preserve and enhance the hydrologies, ecologies and geomorphologies on site. By rising to the eco-revelatory challenge, we were able to take a potentially static bank site and transform it into a thoughtful, sustainable and beautifully functioning landscape.

Indeed, the success of this project is in no small part due to the fact that the client/bank president is an amateur herpetologist and life-long conservationist. He quickly understood the benefits of eco-revelatory design. Months after implementation, he was pleased to inform us that several patrons had moved their accounts to the bank because they were drawn to the beauty of the new landscape. Rather than burying the existing stream for maximum land utilization (i.e. parking) and implementing the conventional pipe-to drain and inlet-to-stream stormwater strategy, the 'new landscape' kept the existing landscape in tact as much as possible. By incorporating a natural above-ground, stormwater attenuation system and integrating smaller bioretention areas to filter and remediate pollutants, we were able to preserve and revitalize the existing urban stream on site (Figure 12.2). We also employed a similar remediation strategy to that of Peter Latz in his Duisburg Nord project, in that we implemented hydric species plants for pollutant uptake within the biofilters (Figure 12.3). The fact that average people would actually move their accounts to this bank because of their experience on-site is a testament to the success of the bank's unique 'eco-revelatory' landscape.

If only all projects could be this simple. Having to convince clients to spend a little extra in the early design stages to work out the intricacies of some of these systems is not always an easy task. Sometimes they simply do not have the design budget or willingness to take on an alternative approach. They are reluctant to depart from convention even if it would save them considerable money during the construction process. While it is fairly straightforward to clear, grub, flatten and bury everything on site for ease of construction, there is more complexity involved in lying lightly on the land, working with the elements on site and implementing systems that complement the surrounding environment in a mindful way.

Figure 12.2 Bank site and Meadow Creek Stream *after* design implementation

Figure 12.3 Commercial bank site as hybrid landscape – engineered biofilters serve as urban gardens

As a seasoned Charlottesville surveyor once told me: 'It costs a little bit to get it done right. But it costs even more to get it done wrong.' Nothing could be truer when dealing with our environment. These words seem to have echoed to the far reaches of the Commonwealth. In recent years, the state of Virginia has begun to take legislative measures to protect its impaired waterways, reservoirs and the Chesapeake Bay. Alternative stormwater management strategies, such as the ones proposed for the bank project, are not only becoming more commonplace but are now required by law in many localities within the state. As more and more people experience these hybrid landscapes, the mission and vision of the eco-revelatory practitioners are closer to being realized.

Notes

1. Quoted in an excerpt in McCormick (2010), p. 46.
2. Until its recent deterioration due to unmanageable sediment loads and algae blooms, the Chesapeake Bay was one of the primary resources for seafood and other domestic products for the Mid-Atlantic United States.

References

Helphand, Kenneth and Robert Melnick (1998) *Landscape Journal* Vol. 17, p. ix (Special Issue, The University of Wisconsin Press).
McCormick, Kathleen (October 2010) *Landscape Architecture Magazine* (Washington, DC: ASLA, Ann Looper Pryor Publisher).
Meyer, Elizabeth (2008) 'Sustaining beauty. The performance of appearance – A manifesto in three parts', *Journal of Landscape Architecture,* 3(1), 6–23.
Van Der Ryn, Sim and Stuart Cowan (1996) *Ecological Design* (Washington, DC: Island Press).

Conclusion and Outlook

Manuela Achilles and Dana Elzey

The challenges we face in environmental degradation, climate change, and resource exhaustion are monumental, existential and collective. They are hard because they are emerging on an unprecedented scale, occur slowly and fitfully, and cannot be addressed by any one answer. In other words, there is no 'one size fits all' solution – each society's response must emerge from and be tailored to fit its unique historical, cultural and political context. At the same time no single nation, no matter how bold and far reaching its response, can envision and build the sustainable societies of the future on its own. It will take international cooperation and coordination, free of the trappings of self-centered diplomacy and the tendencies to let others bear the burden of change while doing nothing, to ensure the sustainability of our natural resources and ecosystems for the generations to come.

Despite the many hurdles confronting us, there is no doubt that with our accumulated knowledge and experience we now conceptually understand the challenges we face better than ever. To cite Garrett Hardin's seminal essay, we have to learn to manage a common resource and avoid 'the tragedy of the commons', that is, the depletion of a shared resource that results when everyone pursues his own self-interest while at the same time having relatively free access to the shared resource.[1] The case of Easter Island, as related by Jared Diamond in *Collapse: How Societies Choose to Fail or Succeed*, is sometimes referred to as a cautionary tale. The case bears some resemblance to several aspects of the current situation affecting the planet. It shows what happens when a human population on the island is driven by cultural beliefs and traditional practices – construction of huge stone memorials to their dead rulers – to deplete a natural resource essential to their survival to the point of no return. In particular, the larger species of trees, used to transport and erect the statues, was also necessary for fuel, heating and cooking, and shelter. The case illustrates how difficult it may be, even for a small population of some 10,000 people living on a tiny island of some 60 square miles, to recognize, accept and take appropriate response to an existential environmental threat.

The contributors to this book are cautiously optimistic that the dire consequences faced by the Easter Islanders can be avoided on a global scale. As Nobel Prize laureate Elinor Ostrom has shown, many communities before us have succeeded in managing a common resource successfully.[2] While by no means exhaustive, the current volume touches on a number of promising, perhaps even critical, steps in addressing the challenges of sustainability. These include the effective economic valuation of ecosystems and resources, the design and implementation of efficient urban living spaces, novel approaches to land reclamation and re-thinking how the built environment can create connection between humans and the natural environment, international collaboration, exchange, and education, as well as grassroots political movements and government interventions such as ecological tax incentives. As a whole and in its individual parts, the volume rests on the premise that we can learn by comparing and contrasting perspectives on sustainability in the United States and Germany. What, then, are some of the book's take-aways and how can they be signposts to a sustainable future? What is the outlook for Germany, with its ambitious and far-reaching plan to achieve sustainability, and for the United States, with its wealth and global influence, yet suffering polarization and paralysis in its response to global climate change?

As discussed by several authors in this volume, Germany currently represents one of the world's foremost experiments in attempting to achieve a future free of carbon-based energy. However, this is not merely an attempt to create a society deriving its energy needs from renewable sources, but one which manages to deliver a high standard of living and maintains Germany's position among the world's top economic powers. As Marrs notes in his article, Germany is attempting to break free from a more than one hundred-year pattern among industrialized nations, where energy is generated by burning fossil fuels to produce steam. Germany's motivation does not arise from altruism, rather a central aim driving the 'energy transformation' (*Energiewende*) is to develop and implement technologies that can be exported and create the jobs of the future.

Among Germany's greatest challenges in achieving the 'energy revolution' are the significant economic costs it will require, the risk of de-industrialization, and the need for massive infrastructure changes needed to adapt to a much more distributed form of power generation. However, despite the costs and the lack of immediate tangible benefits as yet, the citizens of Germany have thus far proven that they have the political will to go forward. They have continued to set (and meet) ambitious goals for greenhouse gas emission reductions, increasing energy efficiency and conservation, and power generation using renewable sources.

Even though Germans are in heated debate regarding the practicalities of this ambitious journey, fraught with risk, they remain steadfast in their rejection of what might be considered the most innovative energy source of the past century, nuclear fission. As Achilles makes clear in her article, the reasons why Germany has come to this decision are bound up with its historical experience, political composition and cultural values. While the Chernobyl and Fukushima disasters have played a role in provoking public outcry against nuclear power, the ultimate cause and effect lies more deeply embedded in the German culture and experience. This reality speaks to both the outlook for the future (i.e. to Germany's likely commitment to the present course) and the importance of contextual factors when planning a response to such complex, interconnected transformations.

As to the question of how Germany has come to be the trial case for a sustainable energy future, one very important factor is the emergence of a strong opposition movement rooted in a radical critique of industrialism, over-reaching capitalism, and militarism. This opposition, tied to the very core of Germany's post-war identity, eventually coalesced into the founding of the Green Party, as detailed in the article by Jungjohann and also by Achilles. The transformation from a heterogeneous, anti-establishment movement into a professional party created challenges as well as chances for Greens. While their entry into the political mainstream alienated segments of their radical base, it allowed for the formation of effective governments with different political partners at the state and federal level. The Greens' ability to play the role of 'kingmaker' in the German political system will ensure the party's role as a continued, viable force in the ecological transformation of Germany's industrial society in the future. This trend is supported by the party's middle-aged, university-educated constituency, which is a growing part of the electorate and has thus far strongly supported the idea of economic growth through renewable energy, sustainability-related technologies, and a 21st-century infrastructure.

Is a militant, activist opposition movement necessary to initiate a society's radical change of course? While radical opposition might appear preferable to the 'catastrophe first' option, the Green Party emerged in the specific context of post-war Germany and thus is hard to replicate in a different setting. The at times turbulent and painful shift of the German Green movement from the political margins to the German political mainstream, on the other hand, does make clear that the necessary transformations on the way to a more sustainable society will not come easily. On the contrary, they will require sacrifice, compromise, persistence, and time. In the end, a democratic transition towards a greener future will only succeed to the extent that radically new ideas find fertile soil in the broader strata of a given society, which will require inter-generational and inter-cultural exchange and learning.

When comparing perspectives on sustainability in the United States and Germany, one is at first struck by the current absence of any comparable, prominent institution, or policy framework focused on sustainability at the national level in the United States. While efforts are underway to develop a framework and tools for integrating sustainability into the Environmental Protection Agency, no coordinated committed platform for the promotion of sustainability practice exists. It is therefore difficult to assign any collective view of sustainability at the national scale in the United States. Seemingly hard-wired differences among opposing political parties have not only prevented real progress in addressing the challenges that lie ahead, but in achieving consensus on what the important challenges actually are.

While segments of the American public do seem to display a sense of urgency regarding the dangers of tyrannical government or a gradual decline in moral values, climate change, potential food shortages, diminishing fresh water resources, soil erosion, and other sustainability-related issues are currently shunned in the public debate and often do not seem to be taken seriously. The comparison with Germany raises the question why this is the case. Geographical and cultural isolation are certainly factors. While globalization has undeniably affected the United States in the last few decades, it is the case that international transactions as a fraction of GDP are still relatively low. Also only about one-third of Americans currently own a passport, which is well up from pre-2007, when it became required to have a passport when traveling to Canada or Mexico. US citizens do not have or seek the frequent opportunities most Europeans have to meet and interact with people of other languages and cultures, and seem thus less aware of such differences and less experienced in dealing with them. Needless to say, the vast size of the United States helps explain in part why more Americans are still perfectly content to limit their travels to the United States: for example, the straight-line distance from Brownsville, Texas, to Los Angeles is greater than from Naples to Oslo.

Related to the size argument is, of course, also the relatively low population density in the United States. Lower population density tends to dilute felt environmental impacts, rendering less visible the stress on resources or on infrastructure which often function as a trigger for social change and as a catalyst for new public policies. While Germany's overall population density is 608 persons per square mile, it is only 78 in the United States[3]. Therefore the situations and forces causing one to confront sustainability issues may be less evident in the daily lives of many Americans.

Also differing is the role of government in identifying and implementing means to enable and accelerate the transition to a sustainable society. Germany's example illustrates the effective power of public policy in incentivizing reform, leading to real changes in human perception and behavior. The United States is certainly capable of dramatic action, such as that occurred in the wake of the launch of the first Russian satellite or the attacks on the World Trade Center. It is no secret that policies in the United States are often more market-based than regulatory, however, while Germany is known to assign relatively strong regulatory powers to the national state.

This collection of essays supports approaches that are in tune with a society's traditions whatever their origins, while also suggesting pathways for improvement and change. Whereas it is true that American environmental and energy policy of recent decades has been lacking a global perspective that draws on international best practices, it is important to remember that the United States was not so long ago at the frontier in environmental policies with its Clean Air and Clean Water Act. As a matter of fact, Germany established a Ministry for Environmental Affairs years after the Environmental Protection Agency was founded in the United States. It is here that cultural cross-fertilization may help. In the same way in which early US environmental policies were a worldwide inspiration in the 1970s, a transatlantic dialogue with Germany might re-open doors that have been shut in previous decades, thus connecting the United States back to its earlier environmental tradition.

Medearis' essay points to the surprisingly rich late 19th- and early 20th-century history of the exchange of knowledge and resources between Germany and the United States. This history encompasses basic scientific discovery, education theory and practice, resource conservation, and green urban planning. The US National Forestry Service and the modern research university, for instance, are innovations imported into the United States from Germany. Seeking to revive and expand the sense of transnational cooperation and exchange, Medearis (and this volume more generally) emphasizes the effective transfer of knowledge and best practices from localities that have demonstrated success in increasing sustainability and the use of renewable energy to those just beginning to grapple with these challenges. Through the example of the Verband Regional Stuttgart and the Northern Virginia Regional Commission, he illustrates a model for the *regional* transfer of knowledge and practice in the areas of energy, transportation and housing. This well-established record of successful regional exchange and collaboration bodes well for the future of renewable energy and sustainability below and beyond the radar of the nation and/or state.

An important trend undercutting the national frame of both analysis and practice is the increasing proportion of the world's population – estimated at nine billion by 2050 – living in cities. It is hard to envision Germany without Berlin, Munich, Cologne or Hamburg. In the same way, one cannot imagine the United States without New York, Chicago, San Francisco or Boston. It is quite possible that cities may help us find an answer to the environmental challenges that we face, in spite of any cultural or historical hurdles that may exist. Beatley, who sees in urbanization the opening of a 'global urban epoch', notes that cities have enormous advantages in meeting the challenges looming ahead. They enable a given number of people to achieve a higher quality of life while consuming resources more efficiently than otherwise possible. Innovative urban design allows much greater efficiency in the use of energy for heating and cooling, but also in meeting needs for individual mobility. Urban living also eases the challenges of organization and economic reform needed to achieve a transition to sustainable society. While the challenges of climate change and sustainability are not driving the process of urbanization, this trend will continue (due to increased opportunities for employment and the promise of improved quality of life). As the planet is called upon to provide an adequate quality of life for an increasing population, highly efficient and dense urban habitats will become an essential feature if we are to achieve sustainability.

The essays by Beatley, Medearis, Kueppers/Maue/Kristan and other authors assembled in this volume suggest that one must look at the local and regional levels, as well as at the 'global' connections to see real progress in sustainability in the United States. Cities and regions pushing ahead with sustainability initiatives and green enterprise are likely to become the nuclei from which the rest of the country can draw inspiration and example, and so gradually transition to sustainability.

Local champions for sustainability should continue their efforts to inform and educate their citizenry, and to organize and build coalitions, both locally and where possible with international partners, such as Germany, at the city or regional level.

A highly localized, cross-fertilizing area of transnational sustainable practice has to do with increasing awareness of the roles played by the natural world, such as disease mitigation, water purification, sanitation, crop fertilization, and many more. Human-built designs, which purposefully integrate, preserve, and reveal such natural processes, thereby allowing us to see and to experience such functionality, help raise our appreciation of the natural world. Such eco-revelatory design (as discussed by Ryang in this volume), in which sustainable technology and practice has been successfully merged with art and design, has important examples in Germany and the United States. This approach is related to the capacity for re-thinking long-standing patterns of land use and resource planning, cited by Siewecke as essential to the transition from a carbon-based energy infrastructure to a renewable energy economy. Sieweke's essay also stresses the importance of preserving a cultural memory of the industrially scared landscape within the context of establishing a more sustainable future.

While advances in the realms of policy, technology, and city design may go far in helping us address the challenges, we cannot achieve a practically sustainable human society while maintaining a high standard of living without changes in human behavior. Such changes will require a reevaluation and adaptation of our core values as well as the emergence of a new, more holistic narrative about how our modern societies see themselves. The essays in this book seek to provide stories, examples, and pragmatic solutions that can be part of such a new narrative that already resonates in Jeremy Rifkin's *The Third Industrial Revolution*, Michael Braungart's and William McDonough's *Cradle to Cradle: Remaking the Way We Make Things*, Ralf Fücks' *Intelligentes Wachsen: Die Grüne Revolution*, or Tim Beatley's *Biophilic Cities: Integrating Nature into Urban Design and Planning*.

The reconsideration of core values and an emerging new narrative that supports highly localized and yet comprehensive sustainability revolutions must include the basic ideas of what constitutes a high quality of life, the definitions of success and wealth, and the nature of our relationship to the natural world. For the United States, it will require reconsideration of the appropriate balance between individual freedoms and the public best interest, assigning appropriate value to natural resources and questioning certain aspects of American 'exceptionalism'. For Germany, it will entail making sure that the forward-looking vision of a sustainable society is not buried beneath the day-to-day financial constraints in the wake of Europe's financial crisis, or gets lost in the sometimes-heard emphasis on adaptation to climate change as outlined by MIT Professor of Environmental Economics Michael Greenstone and others.

A potentially critical mechanism to achieve a more sustainable future within the framework of market-based societies is to assign economic value to the natural goods and services we derive from the ecosystem. Without this, as White explains in his article, we are not in a position to evaluate or appreciate the scarcity of a given resource. This lack of knowledge, in turn, makes it virtually impossible for us to make wise choices when considering how best to use these resources. Furthermore, as White points out, new innovations and economic opportunities will emerge through the valuation of these resources. The necessity to introduce such measures will increase with the rising pressures of population, reduction in resource reserves, and increasing competition for remaining resources.

On a deeper, perhaps philosophical level, the valuation of ecosystem services is about 'recognizing the true cost of our consumptive activities'. Perhaps with greater awareness of both scarcity and the dangers of consumption, we might find the political and social will to question economic growth at any price and to focus on maximizing quality of life or well-being, rather than the accumulation of wealth.

In conclusion, we will need both technological advances and adaptive transitions in human behavior to meet the challenges ahead. Changing deeply ingrained cultural values and perceptions and developing a new narrative of a sustainable society in which we recognize ourselves may be our most difficult challenge. Innovative social, economic, and technology policy will be essential. However, among our most valued resources will continue to be cross-cultural sharing and collaboration, it is here that our own cultural values and the role they play in shaping our lives are revealed in sharpest contrast.

Notes

1. Garrett Hardin, 'The Tragedy of the Commons Science', New Series, Vol. 162, No. 3859 (Dec. 13, 1968), pp. 1243–1248.
2. Elinor Ostrom (1990) *Governing the Commons: The Evolution of Institutions for Collective Action* (Cambridge: Cambridge University Press).
3. http://www.nationmaster.com/graph/geo_pop_den-geography-population-density.

Index

Printed and bound in the United States of America